THE EFFECT OF RIPARIAN ZONES
ON NITRATE REMOVAL BY DENITRIFICATION
AT THE RIVER BASIN SCALE

by

HOANG NGUYEN KHANH LINH

THE EFFECT OF RIPARIAN ZONES ON NITRATE REMOVAL

BY DENITRIFICATION AT THE RIVER BASIN SCALE

DISSERTATION

Submitted in fulfillment of the requirements of
the Board for Doctorates of Delft University of Technology and of
the Academic Board of the UNESCO-IHE Institute for Water Education
for the Degree of DOCTOR
to be defended in public
on Wednesday, 4 December, 2013 at 10:00 hours
in Delft, The Netherlands

by

HOANG Nguyen Khanh Linh

born in Ho Chi Minh city, Vietnam
Master of Science in Hydroinformatics (with distinction)
UNESCO-IHE, Delft, The Netherlands

This dissertation has been approved by the supervisors:
Prof.dr.ir. A.E. Mynett
Prof.dr.ir. A.B.K. van Griensven

Members of the Awarding Committee:

Chairman	Rector Magnificus, Delft University of Technology
Vice-chairman	Rector, UNESCO-IHE
Prof.dr.ir. A.E. Mynett	UNESCO-IHE/Delft University of Technology, supervisor
Prof.dr.ir. A.B.K. van Griensven	UNESCO-IHE/VU Brussel, supervisor
Prof.dr.ir. H.H.G. Savenije	Delft University of Technology
Prof.dr.ir. M.E. McClain	UNESCO-IHE/Delft University of Technology
Prof.dr.ir. D.P.L. Rousseau	Ghent University, Belgium
Prof.dr. N. Fohrer	University of Kiel, Germany
Prof.dr.ir. N.C. van de Giesen	Delft University of Technology, reserve member

CRC Press/Balkema is an imprint of the Taylor & Francis Group, an informa business

Published by:
CRC Press/Balkema
PO Box 11320, 2301 EH Leiden, The Netherlands
e-mail: Pub.NL@taylorandfrancis.com
www.crcpress.com - www.taylorandfrancis.com

ISBN: 978-1-138-02405-2

To my dear family

ACKNOWLEDGEMENT

Delft has been my second home for the past 6 years while I worked towards my MSc and PhD degrees. Undoubtedly, this is the most important and meaningful chapter of my life when I learned the most, enjoyed the most, fell down sometimes but then grew up more and more. This is the chapter that I will treasure and remember for my whole life. Before closing this chapter, I would like to express my sincere gratitude to all the people, related or not related to my PhD work, who gave me help and support to finish my research.

First of all, I would like to convey my sincere gratitude to my promoter, Prof. Arthur Mynett, who played the most important role to make this PhD finalisation come true. Thank you very much not only for your continuous support but also for your push at the final stages. Although I felt really stressed and exhausted sometimes by the pressure, I know I could not have finished my PhD without it. I thank you for always believing in me and encouraging me by saying 'Linh, you can do it', and then I also told myself 'Yes, I will try as much as I can'. I am really grateful to you for giving me the chance to attend the IAHR World Congress 2013 in Chengdu, China with you and present our paper. And finally, thank you for all the discussions with you that helped me to find some light in the darkness of this PhD road and feel more confident.

Secondly, I would like to express my gratitude to my co-promoter, Prof. Ann van Griensven, who was my supervisor in both Master and PhD theses. I am truly grateful to you for giving me a chance to continue my PhD following the AQUAREHAB project. Thank you for your guidance, suggestions and help throughout my PhD, especially on SWAT as the main tool for my PhD.

I would like to send my deep appreciation to Prof. Jens Christian Refgaards who always gave me very useful suggestions and ideas for my PhD research. It was my great pleasure to work with you and have very long discussions with you every time I came to Copenhagen. Although it is really a pity that you cannot be present in my defence, I would like to send my great gratitude for your warm welcome, your kind help and your precious time spent with me.

I also give my sincere gratitude to Mr. Bertel Nilsson for his great help during my time in GEUS and in Copenhagen. Thanks for finding a very good accommodation for me, arranging me an office, arranging the key, coming to say hello every morning and helping me with all the data and connections. Your gentle care made me feel warm and welcomed in GEUS and I really enjoyed that time a lot. I also thank Dr. Lars Troldborg for providing me data, Ms. Heidi Barlebo for your warm welcome to the Hydrology group and my friends in Copenhagen: Xin, Xiulan, Kinza for spending enjoyable time with me during lunch and after working hours.

I want to send my special thanks to my landlords Inger and Hans Ole Hansen for providing me such a wonderful accommodation that made me feel like home in Copenhagen. Inger, I cannot express enough my thanks to you for always caring about me in everything I did and your comfort in one of the most terrible moments of my life. I really appreciate your kindness to let me stay in your lovely house the second time when I really needed a private place to work.

My genuine thanks are sent to Sha, my best friend during my PhD. It is such a pleasure to have you beside me in the last 3 years. Thanks for spending enjoyable time with me, considering me as 'automatically invited guest' for all your special dinners, helping me every time I need, and being available for me to disturb whenever I have problems. Without you, my PhD life would have been very boring.

I would like to thank my closest colleague Girma, who went along with me for the whole 4-year PhD period. Being in the same project and having the same supervisors brought us close to each other although we are from different countries. Thanks for sharing with me all the difficulties and hard times in this PhD life and having a talk with me everyday. I will not forget the many weekends that we had to spend in TU library to finish our PhD drafts. Thanks for accompanying me during this difficult and stressful time.

My thanks also go to my international friends: Fiona, Yuli, Anuar, Isnaeni, Yos, Sony, Leo for creating such a friendly atmosphere for me to join your Indonesian-Malaysian community. Thanks to Off, Pim, Nat, and Alida to welcome me to your Thai group. Thanks to Dr Suryadi for your support in my life and my PhD, your gentle care during my time in Delft and the time when we were in Chengdu, and countless number of your dinner invitations. I genuinely feel pleasure to have met and spent such lovely time with all of you.

Thanks to my Vietnamese friends in Delft: be Tam, Truong, Khai, Trang, Minh, Vinh, Thang, anh Tu, anh Ngan for sharing good times with me during our gathering which made my life in Delft more lively. Thanks anh Tu and Thang for joining me for skating practice which helped to reduce my stress in the final time of my thesis writing. And thanks to all my friends in Vietnam: my cousin Dao Chi, Phong, Duy, Chau, co Le for all your support and sharing.

Then, I would like to send my deepest and greatest gratitude to the most important persons in my life.

First, thanks to my Osin for always being beside me for such a long time and sharing with me everything happening in our lives. Although we live far from each other, I feel that you are always there whenever I need you. Thanks for being such an important part of my life that I would like to treasure it for my whole life. Your support, encouragement and funny conversations never failed to make me laugh when I was in difficult times. Whatever happens, you are always the most special person in my life.

Secondly, thanks to Thuy Phuong for being my best friend for 15 years. Sorry for being away from you for such a long time although you asked me to come back soon. I cannot express how grateful I am to have you in my life. Although we are not in close contact, our friendship has never changed. Thanks for always being available for me whenever I need you to lift me up.

Finally, I would like to send my greatest gratitude to my family: my parents and my younger brother for their unconditional love and endless belief and support for me during my whole life. I am really grateful to have such good family who wish me the best for my life, always welcome me to return anytime I want and support me in everything I want to do. Thanks to my younger brother to be beside my parents to take care and entertain my parents during my absence. This thesis is dedicated to my family as my most genuine gratefulness to their love and support.

Delft, October 2013

Linh Hoang

SUMMARY

The riparian zone, the interface between terrestrial and aquatic ecosystems, plays an important role in nitrogen removal in spite of the minor proportion of the land area that it covers. This is verified in a large number of studies related to the effect of wetlands/riparian zones on the discharge of nutrients. The nitrogen removal in riparian zones is mostly caused by denitrification, which is favoured by anaerobic condition created by high water level in riparian zones. Most of the studies related to the effect of riparian zones on nitrogen removal are limited to small scales and field work. Very limited research has been carried out in modelling their effects at large scales or river basin scales.

Nowadays, there are many river-basin-scale models available that are able to deal with predictions of pollutant loading from diffuse sources. Several wetland/riparian models are available for simulating hydrological and chemical processes in wetlands/riparian zones. However, there are limited studies on integrating wetland/riparian zone models in river basin modelling for evaluating the effect of wetlands/riparian zones at river basin scales. The SWAT model is a well-known and broadly used model to simulate hydrological and nutrient transport processes. In terms of riparian zone modelling, there are very few studies in this field, although SWAT does have a sub-module to estimate flow and pollutant retention in buffer strips based on empirical equations which are derived from observations.

The main objective of this thesis is to evaluate the effect of riparian zones on nitrate removal at the river basin scale using the SWAT model. This thesis focused on modifying the SWAT model by (i) adding the landscape routing concept across different landscape units and (ii) adding the Riparian Nitrogen Model (RNM) in SWAT to simulate the denitrification process in riparian zones. The first modification aims at taking into account the landscape position of a Hydrological Response Unit (HRU) and creating a relationship between HRUs in upland and in lowland landscape units which is considered important for flow and pollutant transport processes. Based on the first modification, the second modification introduces a conceptual model called Riparian Nitrogen Model into SWAT for representing denitrification process in HRUs that belong to riparian zones. In this model, the denitrification rate is assumed to decline with depth and the denitrification process is activated below the groundwater table. This interaction occurs via two mechanisms: (i) groundwater that passes through the riparian buffer before discharging into the stream and (ii) surface water which is temporarily stored within the riparian soils during a flood event. With these modifications, this thesis is contributing to further developing and improving the SWAT model. The SWAT modifications were tested with a simple hypothetical case study and applied to the Odense river basin, which is an agricultural-dominated area and a densely tile-drained river basin in Denmark.

The main findings from this thesis can be summarised as follows:

- The SWAT model performed well in replicating the daily streamflow hydrograph at the calibrated station although some of the peak flows were under- or over-predicted and variations in low flows were not captured well. In terms of nitrogen simulation, the SWAT predicted results compared well in magnitude and in variation of the predicted versus measured daily nitrogen fluxes, which implies that SWAT can be an effective tool for simulating nitrogen loadings in the Odense river basin.

- Compared to the existing DAISY-MIKE SHE model of the Odense river basin, taking into account the uncertainty of soil hydraulic properties and slurry parameters, the SWAT flow and nitrogen results fitted quite well within the uncertainty ranges of the DAISY-MIKE SHE model. By comparing the annual average water balances between the two models, a striking water balance difference was observed, which shows that virtually all of the subsurface flow occurred as tile flow in DAISY-MIKE SHE while the actual subsurface flow inputs were roughly split evenly between tile flow and groundwater flow in SWAT. In terms of nitrogen fluxes, both models predicted that nitrate from tile flow was the dominant source.

- A comparison was implemented between different SWAT setups including (i) SWAT without tile drainage, (ii) SWAT with tile drainage. It is known that the Odense river basin is a densely tile-drained area and the SWAT model without tile drainage is not realistic since it neglects the importance of the tile drainage in the system. It was indeed observed that including tile drainage significantly changed the flow pattern, implying that tile drainage has an important contribution to the streamflow and is a very important component that should be included in the model of the Odense river basin. However, a surprising result was obtained in that the SWAT setup without tile drainage simulation achieved a much better fit to the measurements (very high Nash Sutcliffe coefficient) than the SWAT setup with tile drainage, even after autocalibration. From this it can be concluded that statistical metrics, which are useful for evaluating the goodness-of-fit of simulated results versus measured data, are unable to fully distinguish whether a model is capable of representing water and pollutant pathways correctly. Therefore, it is extremely important to select a suitable model structure for a particular case study, as much as possible based on information about the area, expert knowledge, observations from the field, etc.

- The approach to represent the landscape variability in SWAT (SWAT_LS) gave rise to two modifications to the SWAT model: (i) dividing a sub-basin into two landscape units: upland and lowland, and (ii) allowing hydrological routing between them. The results showed that the sensitivities of flow-related parameters on flow response in SWAT_LS were similar to the original SWAT2005 model. The flow value for each flow component in SWAT_LS had a small change compared to the original SWAT2005. Consequently, it can be concluded that the added routing between landscape units can affect the volume of flow, but does not influence the flow pattern within the sub-basin. In comparison with SWAT2005, SWAT_LS decreases the surface runoff because part of the upland surface runoff is infiltrated into the lowland groundwater, decreasing the overall surface runoff because of the longer lag-time for groundwater flow to move across two landscape units before reaching the river. Moreover, the proportion between upland and lowland areas seems to have a very strong effect on the upland flows; however, the effect on flows from the whole sub-basin is not significant. With this approach, it is possible to take into account the landscape position of each HRU, and therefore it is possible to evaluate flow and nitrogen results for each landscape unit. In addition, this approach also represents the interaction between upslope HRUs and downslope HRUs, which gives a better representation of the hydrological processes in reality.

- The Riparian Nitrogen Model brings a new concept to modelling the denitrification process in riparian zones in the SWAT model. With the added landscape approach (SWAT_LS) it is possible to differentiate among HRUs that belong to riparian zones, and therefore it is possible to apply the Riparian Nitrogen model only in the riparian HRUs. When testing SWAT_LS in the hypothetical case study, the riparian zone was seen not to have any effect when deep groundwater flow or surface runoff dominates. Denitrification in the riparian zone mainly occurs when the riparian zone receives high amounts of tile flow, which bring high amounts of nitrate and cause a rise in the perched groundwater table in the riparian zone. Contrary to the original SWAT2005 model, the SWAT_LS model is able to evaluate the efficiency of the riparian zone with respect to nitrate removal by denitrification at the river basin scale.

- The application of the modified SWAT in the Odense river basin showed that, compared to the original SWAT2005, SWAT_LS gave improvements in the simulation of flow and nitrate fluxes as indicated by the Nash-Sutcliffe coefficients. A comparison between the two SWAT versions, taking into account parameter uncertainty by running 5000 Monte-Carlo simulations, showed that SWAT_LS had a much higher number of parameter sets that gave satisfactory performances (behavioural models) in both daily and monthly time steps. This implies that SWAT_LS has a higher probability of achieving a satisfactory representation of the modelled river basin. Although there is a big difference between the number of behavioural models, the uncertainty bounds are compatible between the two versions. It was observed that in the Odense river basin the riparian zones do not have a significant effect on nitrate removal because a large part of the riparian zone is dominated by tile drainage.

- Because of limited availability of measurements on denitrification and the assumption that all riparian zones in the Odense river basin have the same characteristics, the estimation of denitrification only took into account the denitrification-related parameter uncertainty in (i) the current situation and (ii) a hypothetical condition when all riparian zones are not drained. The results showed that the uncertainty band of scenario (ii) is broader than the uncertainty band of the present condition, because denitrification occurs in a larger area. In the present condition, the nitrate removal is only about 4~17% taking into account uncertainty. However, if all riparian zones in the area are not drained and can really perform their retention function, the effectiveness of the riparian zones on nitrate removal will increase dramatically to about 25~85%.

SAMENVATTING

De oeverzone, het grensgebied tussen land en water, speelt een belangrijke rol bij het verwijderen van stikstof uit ecosystemen, ook al is het gebied in oppervlakte niet zo groot. Een groot aantal studies naar het effect van oeverzones op de afvoer van nutriënten bevestigt dit. Om stikstof te verwijderen zijn anaerobische condities gewenst en deze doen zich voor tijdens een hoge waterstand. De meeste studies naar denitrificatie in oeverzones beperken zich tot veldonderzoek op kleine schaal. Er is nog zeer weinig onderzoek gedaan naar het modelleren van deze processen op stroomgebied schaal.

Wel zijn er tegenwoordig veel computermodellen beschikbaar om de (grond)water stroming te modelleren, inclusief het transport van verontreiniging. Er zijn modellen die hydrologische processen simuleren, of chemische processen. Maar er zijn weinig modellen die een geïntegreerde benadering volgen op stroomgebied schaal. Het SWAT model is een bekend en wijd verbreid Open Source model dat zowel hydrologische processen als het transport van nutriënten kan modelleren. Voor het modelleren van oeverzones zijn weinig veldstudies uitgevoerd, alhoewel SWAT wel een sub-module bevat om de grondwater stroming en afbraak van verontreiniging te onderzoeken in buffer strips, maar gebaseerd op empirische relaties verkregen uit lokale waarnemingen.

Het hoofddoel van dit proefschrift is om het effect na te gaan van oeverzones op het verwijderen van stikstof op stroomgebied schaal, gebruik makend van het SWAT model. Het onderzoek is erop gericht om het SWAT model uit te breiden door (i) een routine toe te voegen die de stroming tussen verschillende landschapscomponenten kan weergeven, en (ii) een Riparian Nitrogen Model (RNM) aan SWAT toe te voegen dat het denitrificatieproces in oeverzones representeert.

De eerste uitbreiding is erop gericht om de specifieke verdeling van Hydrologische Respons Units (HRUs) mee te nemen en een verband te creëren tussen HRUs bovenstrooms en benedenstrooms, aangezien dit van belang is voor het transport van verontreinigingen. De tweede uitbreiding betreft het ontwikkelen van een conceptueel model genaamd Riparian Nitrogen Model dat vervolgens in SWAT kan worden ingebracht om het denitrificatieproces in HRUs nabij de oeverzones te kunnen weergeven. In dit model wordt aangenomen dat de snelheid van denitrificatie terug loopt met de diepte van de grondwaterspiegel. De uitwisseling gebeurt op twee manieren: (i) via het grondwater dat door de bufferzone stroomt; en (ii) via het oppervlaktewater op de oeverzone in geval van overstroming. Dit proefschrift tracht bij te dragen aan de verdere ontwikkeling van SWAT door deze uitbreidingen eerst te testen in een hypothetische situatie en vervolgens in het Odense stroomgebied, een dichte lappendeken van landbouwgebiedjes in Noord Denemarken.

De belangrijkste conclusies van dit proefschrift kunnen als volgt worden samengevat:

- Het SWAT model kan het stromingsgedrag goed weergeven op tijdschalen van een dag, hoewel variaties in piek afvoeren niet goed konden worden gerepresenteerd.

- Het stikstofgehalte voorspeld door SWAT kwam redelijk overeen qua niveau en ruimtelijke verdeling met gemeten dagelijkse waarden, wat impliceert dat SWAT een goed instrument kan zijn om de belasting van het stroomgebied van de Odense te kunnen simuleren.

- Vergeleken met het bestaande DAISY-MIKE SHE model van het Odense stroomgebied komen de SWAT resultaten voor stroomsnelheden en stikstof concentraties goed overeen, gelet op de onzekerheden die een rol spelen. De waarden voor de waterbalans op jaarbasis gaven wel een groot verschil: in DAISY-MIKE SHE vond vrijwel alle afstroming via drainage buizen plaats, terwijl de afstroming in SWAT ongeveer gelijk verdeeld was tussen drainage buizen en grondwater. Beide modellen voorspelden wel dezelfde afname aan stikstof gehalte.

- Er is een vergelijking gemaakt tussen verschillende modelconfiguraties van SWAT, (i) zonder drainage buizen, (ii) met drainage buizen. Het is namelijk bekend dat het Odense stroomgebied een dichtgepakt gebied met drainage buizen is en een SWAT model dat dat niet meeneemt lijkt niet realistisch. Uit een onderling vergelijk bleek dat de configuratie met drainage buizen tot een aanzienlijke andere afstroming leidde, wat het belang aangeeft van het meenemen van drainage buizen in het model van het Odense stroomgebied. Het was daarom verrassend om te constateren dat het SWAT model zonder drainage buizen veel beter in overeenstemming was met de metingen (zeer hoge Nash Sutcliffe coëfficiënt) dan SWAT met drainage buizen, zelfs na auto-calibratie. Daaruit kan worden geconcludeerd dat statistische technieken die zeer geschikt zijn om de relatie tussen voorspellingen en waarnemingen vast te stellen, niet altijd in staat blijken om het stromingspatroon en de verspreiding van verontreinigingen correct te beoordelen. Het blijft dan ook van belang om de juiste modelconfiguratie te kiezen voor een bepaald gebied, zoveel mogelijk uitgaande van specifieke gebiedsinformatie, kennis van experts, veldwaarnemingen, etc.

- Het opzetten van een modelconfiguratie met landschapsvariatie (SWAT_LS) vereiste twee aanpassingen van het standaard SWAT model: (i) onderscheid tussen bovenstroomse en benedenstroomse gebieden, en (ii) onderlinge uitwisseling van stroming tussen beide gebieden. De resultaten lieten zien dat het stroombeeld van SWAT_LS en SWAT2005 vergelijkbaar was. De waarden in SWAT_LS waren weliswaar iets verschillend, maar over het algemeen bestond er redelijke overeenstemming. Daaruit kan worden opgemaakt dat de uitwisseling tussen deelgebieden weliswaar invloed heeft op het totale debiet, maar niet op het stromingspatroon binnen de deelgebieden.

- Vergeleken met SWAT2005 is de afstroming aan het oppervlak in SWAT_LS minder omdat een deel van de bovenstroomse afstroming als grondwater verder stroomt in het benedenstroomse deel. Afstroming via het grondwater verloopt namelijk veel trager. Daar komt bij dat de verdeling tussen bovenstroomse en benedenstroomse deelgebieden van grote invloed blijkt op het bovenstroomse deel, maar niet op het totale stroomgebied. De aanpak ontwikkeld in dit proefschrift maakt het mogelijk om rekening te houden met de specifieke locatie van iedere HRU waardoor het mogelijk wordt om het stroombeeld en stikstof afbraak voor ieder deelgebied te bepalen. Bovendien kan de uitwisseling tussen bovenstroomse HRUs en benedenstroomse HRUs worden weergegeven, wat beter overeenkomt met de feitelijke hydrologische processen in het gebied.

- Het Riparian Nitrogen Model (RNM) is een nieuwe bijdrage aan het modelleren van denitrificatie in oeverzones met behulp van SWAT. De landschapsbenadering in SWAT_LS maakt het mogelijk om onderscheid te maken tussen HRUs met of zonder oeverzones en om RNM alleen te gebruiken in gebieden *met* oeverzones. Tijdens het testen van SWAT_LS aan de hand van een hypothetische casus bleek dat het al dan niet meenemen van de oeverzone geen enkele invloed had in geval van diepe grondwaterstroming of afstroming via de oppervlakte. Denitrificatie gebeurt voornamelijk als de oeverzones veel drainagewater ontvangen met hoge concentraties nitraat die vervolgens worden opgenomen in het grondwater. In tegenstelling tot SWAT2005 is SWAT_LS wel in staat om de invloed van de oeverzone te bepalen op het verwijderen van nitraat in het stroomgebied.

- Het gebruik van de aangepaste versie van SWAT in het stroomgebied van de Odense laat zien dat verbeteringen worden bereikt bij het simuleren van het stroomgedrag en nitraat verspreiding, zoals terug te vinden in de Nash-Sutcliffe coëfficiënten. Een vergelijk tussen de twee SWAT versie gebaseerd op 5000 Monte-Carlo simulaties om onzekerheden in parameters vast te stellen, laat zien dat SWAT_LS een veel groter aantal parameter combinaties had die voldeden, zowel op tijdschaal van dagen als van maanden. Dit impliceert dat SWAT_LS beter in staat moet worden geacht om het stroomgebied juist te modelleren. Ondanks dit verschil in aantal zijn de onzekerheidsbanden van beide modellen nagenoeg gelijk. In het stroomgebied van de Odense dragen oeverzones minder bij aan het verwijderen van nitraat, aangezien een groot deel afstroomt via drainage buizen.

- Vanwege de beperkte beschikbaarheid van metingen en de aanname dat alle oeverzones in het stroomgebied van de Odense dezelfde eigenschappen hebben kon de schatting van denitrificatie alleen met de nodige onzekerheid worden bepaald: (i) in de huidige situatie, en (ii) in het hypothetische geval dat alle oeverzones niet gedraineerd zijn. De resultaten laten zien dat de bandbreedte in dat geval groter is. In de huidige situatie bedraagt de nitraat vermindering ongeveer 4~17%; als alle oeverzones niet gedraineerd zouden zijn neemt dit effect in belangrijke mate toe, tot 25~85%.

TABLE OF CONTENTS

Chapter 1

INTRODUCTION

1.1 BACKGROUND

1.1.1 River basin-scale models for diffuse pollution modelling and the SWAT model

The EU Water Framework Directive (EC, 2000) has introduced a new approach in water resources protection in which there is a change from focusing on the control of point sources of pollution to integrated pollution prevention at river basin level and setting water quality objectives for the entire basin. This new policy requires the integration of all water quality issues, related to both point and diffuse pollution sources, at the river basin scale.

Diffuse pollution, especially from agricultural activities, has become a major concern due to past and present efforts in wastewater treatment for industries and households. Compared to point sources, diffuse pollution is more difficult to be controlled since it is characterised by numerous and dispersed sources and the difficulties in tracing its pathways (Yang and Wang, 2010). The application of large amounts of mineral and organic fertilizers in intensely cultivated agricultural areas contributes to excessive environmental loads on soil, groundwater and surface water bodies which affect negatively the biodiversity and human health (Bergström and Brink, 1986; Horrigan et al., 2002; Line et al., 2002).

River basin-scale models, which are capable of estimating pollutant loads from diffuse sources in a basin to the receiving river system, are necessary components of sustainable environmental management for better implementation of the EU Water Framework Directive. Recent reviews by Borah and Bera (2003), Yang and Wang (2010) and Daniel et al. (2011) describe several well-known and operational modelling tools that are able to handle non-point source pollution at the river basin scale. Two of the more widely used of these modelling packages are the Soil and Water Assessment Tool (SWAT) model (Arnold et al., 1998; Arnold and Fohrer, 2005; Gassman et al., 2007) and the MIKE SHE model (Refsgaard and Storm, 1995; Refsgaard et al., 2010), which was developed from the earlier SHE (Système Hydrologique Européen or European Hydrological System) model.

SWAT has been applied worldwide across a wide range of river basin scales and conditions for a variety of hydrologic and environmental problems, as documented in reviews by Gassman et al. (2007; 2010), Douglas-Mankin et al. (2010), and Tuppad et al. (2011). MIKE-SHE is considered to be one of the most comprehensive river basin models developed to date and has also been extensively used for a broad spectrum of hydrologic and water quality assessments in many different regions worldwide as described by Refsgaard et al. (2010) and Daniel et al. (2011).

Borah and Bera (2003) assessed that SWAT is a promising model for continuous simulations in predominantly agricultural river basins. Shepherd et al. (1999) also found that SWAT was the most suitable tool for modelling river basin scale nutrient transport to watercourses in the U.K. A significant number of studies have been carried out to use SWAT to calculate nutrient loads, such as those reviewed in Gassman et al. (2007), Douglas-Mankin et al. (2010), and Tuppad et al. (2011). Numerous SWAT studies also suggest measures for improving water quality based on different management scenarios (Tuppad et al., 2010; Ullrich and Volk, 2009; Volk et al., 2009; Yang et al., 2011; Yang et al., 2009).

SWAT is also able to simulate flow and nutrient fluxes through subsurface tile drains by the subsurface tile drainage component added by Arnold and Fohrer (2005) which was then modified by Du et al. (2005; 2006) and Green et al. (2006). Numerous studies have since been published that describe applications of the SWAT subsurface tile drainage routine, including several that report successful replication of measured streamflow and nitrate levels such as Schilling and Wolter (2009) for the Des Moines River basin in North central Iowa, Sui and Frankengerger (2008) for the Sugar Creek River basin in Indiana, and Lam et al. (2011) for the Kielstau River basin in northern Germany. Due to the wide application of SWAT in densely agricultural areas and its capacity to simulate flow and nutrient fluxes through tile drains, SWAT is chosen to apply in the Odense river basin, Denmark which is the main case study area of this thesis. The land use in the Odense river basin is dominated by agricultural production which results in a densely subsurface tile drain system covering the whole area.

1.1.2 Riparian zones and its modelling at river basin scale

A riparian zone generally encompasses the vegetated strip of land that extends along streams and rivers and is therefore the interface between terrestrial and aquatic ecosystems (Gregory et al., 1991; Martin et al., 1999). This location, between upland and aquatic ecosystems, provides riparian zones the capacity of modifying effects on the aquatic environment. The importance of a riparian zone in a landscape exceeds the minor proportion of the land area that it covers (Gregory et al., 1991). Vegetation in riparian zones can help to intercept solar radiation and lower stream temperature (Gregory et al., 1991) and is also an important source of organic and inorganic material through particulate terrestrial inputs (Roth et al., 1996). Riparian zones are also able to trap sediments amassed in upslope areas (Daniels and Gilliam, 1996).

Interest in riparian zones has focused on the ability to maintain and/or improve water chemistry and the riparian buffer zone has become of critical interest in agricultural settings, where farm management practices have become increasingly intensified (Martin et al., 1999). Flooding of the riparian zone affects the soil chemistry by producing anaerobic conditions, importing and removing organic matters, and replenishing mineral nutrients. The riparian ecosystem acts as a nutrient sink for lateral runoff and groundwater flow from uplands and as a nutrient transformer for upstream-downstream flows (Mitsch and Gosselink, 2000).

There have been a large number of studies on the effect of riparian buffer zones on the discharge of nutrients, particularly nitrate, into fresh water systems. Low concentrations of nitrate have been reported in riparian-zone groundwater, not only in undisturbed headwater watersheds (Campbell et al., 2000; McDowell et al., 1992; Sueker et al., 2001) but also in agricultural watersheds (Hill, 1996; Jordan et al., 1993). Based on data from several papers, Hill (1996) calculated the percentage of removal of nitrate in groundwater passing through the riparian zone in 20 watersheds by comparing the nitrate concentration of groundwater up-gradient from the riparian zone with that of groundwater at the riparian zone/stream interface. He found that in 14 riparian zones nitrate removal was greater than 90% and that nitrate removal in all 20 watersheds ranged from 65% to 100%.

Most of the studies on the nitrate removal capacity of riparian zones are limited in observations at field scales. However, there are several models that are available to simulate nutrient processes in riparian zones. In SWAT, White and Arnold (2009) developed a Vegetative Filter Strips (VFS) sub-model to simulate runoff, sediment and nutrient retention in buffer strips based on a combination of measured data derived from literature and Vegetative Filter Strip Model (VFSMOD) (Muñoz-Carpena et al., 1999) simulations. Wetlands Water Quality Model (WWQM) aims at evaluating nitrogen, phosphorus, and sediments retention from a constructed wetland system (Chavan and Dennett, 2008). Kazezyılmaz-Alhan et al. (2007) developed a general comprehensive wetland model Wetland Solute Transport Dynamics (WETSAND) that has both surface flow and solute transport components.

The Riparian Nitrogen Model (RNM) (Rassam et al., 2008) is a conceptual model that estimates the removal of nitrate as a result of denitrification, which is one of the major processes that lead to the permanent removal of nitrate from shallow groundwater during interaction with riparian soils. Despite the availability of wetland/riparian zone models, there are few models that can evaluate the effect of riparian zones at the river basin scale. One example of consideration for wetland/riparian zone modelling at river basin scale is the study from Hattermann et al. (2006) who integrated wetlands and riparian zones in river basin modelling by adding an equation to simulate nutrient retention in the subsurface and groundwater flow.

1.2 MOTIVATION OF THE THESIS

In this thesis, a model is built for the Odense river basin for flow and nitrogen simulation using the SWAT model suite. Due to its broad applications and the availability of many sub-modules, SWAT is expected to be able to give a good representation for the Odense river basin. The SWAT model is then compared with the DAISY-MIKE SHE model which was already built for the Odense river basin (Hansen et al., 2009). The comparison of these two models has not been carried out previously in terms of evaluation of flow and nitrogen components. This comparison aims at assessing the suitability of the different approaches used in the two models for simulating flow and nitrogen fluxes originating from the Odense river basin.

While the SWAT model is able to give reasonable results in flow and nitrogen fluxes which are shown in many SWAT applications, one of the shortcomings of SWAT is that it does not take into account the effect of the landscape position of the modelling Hydrological Response Unit HRU, and therefore there is no interaction between upland HRUs and lowland HRUs. This also results in the inability to evaluate the effect of riparian zones on flow and nitrogen retention based on their locations between upland areas and streams. It is noted that SWAT does have a sub-module called Vegetative Filter Strips (VFS) to simulate flow and nitrogen retention in buffer strips; however, the sub-module is only limited to estimating retention efficiency based on the relationship with the width of riparian zones.

Therefore, in this thesis, we introduced an approach to SWAT that can take into account landscape variability and allow flow and nitrogen routing between different landscape units. With this approach, it is possible to separate between HRUs in riparian zones and HRUs in upland areas and simulate the interaction in flow and nitrogen fluxes between upland areas and riparian zones. At the same time, it is also possible to evaluate flow and nitrogen retention capacity in riparian zones.

In this thesis, we also added a conceptual Riparian Nitrogen Model (RNM) to the SWAT model for denitrification in riparian zones. The denitrification in this model occurs when groundwater and surface waters interact with riparian buffers. This interaction occurs via two mechanisms: (i) groundwater passing through the riparian buffer before discharging into the stream; and (ii) surface water being temporarily stored within the riparian soils during flood event. The RNM model was used to replace the sub-module for Vegetative Filter Strips in SWAT.

1.3   RESEARCH QUESTIONS

From the motivation of the thesis, the following research questions arise:

-   What is the performance of the SWAT model on flow and nitrogen simulations for a highly tile-drained river basin like the Odense river basin in this thesis?

-   What are the most important processes for flow and nitrogen in the Odense river basin?

-   How different are the performances of the SWAT model and DAISY-MIKE SHE model in flow and nitrogen fluxes? Which one gives better results?

-   How important is the model structure for a good representation of a real case study? Can different models with different structures get good fits to the measured data after calibration? How can we conclude that a model is correct and can reflect reality?

-   At present the SWAT model does not take into account the interaction between HRUs in upland and in lowland areas. If we define landscape positions for HRUs and allow routing of flow and pollution fluxes across different landscape elements from the furthest to the nearest to the streams, will this help to improve the accuracy of the model and will it change hydrological behaviour and water quality processes in the model?

-   What is the effect of the Riparian Nitrogen Model in modelling denitrification in riparian zones when it is added as a sub-module in SWAT? How sensitive are the parameters in the Riparian Nitrogen Model to the predicted nitrate removal efficiency due to denitrification?

- What is the effect of the modified SWAT model, which takes into account landscape variability and uses the Riparian Nitrogen Model to simulate denitrification in riparian zones, on flow and nitrogen simulation? Will the modified SWAT model give a better representation of the Odense river basin? What is the effect of riparian zones in nitrate removal by denitrification in the Odense river basin?

1.4   RESEARCH OBJECTIVES

The main objective of this study is to evaluate the effect of riparian zones for nitrate removal at the river basin scale using the SWAT model. Presently, SWAT is able to estimate nitrate removal in riparian zones using empirical equations that are based on limited observations from literature. Moreover, the present approach of SWAT does not take into account the landscape position of HRUs, thus it is not possible to evaluate HRUs in a certain location of the modelled river basin. To obtain the main objectives, modifications were made in the SWAT model which include (i) adding the landscape identification in HRUs and routing processes across different landscape units and (ii) adding the Riparian Nitrogen Model as a sub-module in SWAT to simulate denitrification process in riparian zones. This study is expected to give a contribution to SWAT development and improvement in flow and nitrogen simulations.

Based on the main objective and research questions, the following detailed objectives are proposed:

Objective 1: Build a river basin scale model of Odense river basin for simulating hydrology and nitrogen transport and transformation using SWAT. Evaluate the performance of SWAT in modelling water quantity and water quality (nitrate in this study) by comparison to observations.

Objective 2: Compare the SWAT model and the existing DAISY-MIKE SHE model of Odense river basin in flow and nitrogen simulations

The DAISY-MIKE SHE model was already built for the Odense river basin. In addition to evaluating the SWAT model based on measured data, the SWAT model was also evaluated by comparing with the DAISY-MIKE SHE model on the distribution of flow and nitrogen components which could not been shown in observations. The comparison between these two models also shows the differences in the performance of a comprehensive fully distributed physics-based model like DAISY-MIKE SHE compared to a simpler semi-distributed conceptual model like SWAT.

Objective 3: Compare between different SWAT models with different model structures

Two different SWAT models with different model structures are to be built for the Odense river basin including: a model without tile drain applied, a model with tile drain applied. The comparison between the two models aims at (i) evaluating the importance of tile drainage process for the Odense river basin and (ii) assessing if calibration can compensate the lacking process with another process to be able to get good fit to observations.

Objective 4. Introduce an approach that takes into account landscape variability in the SWAT model (the modified model is called SWAT_LS). Evaluate the effect of landscape routing in flow and nitrogen simulations by comparing between the SWAT_LS model and the original SWAT model applied in a very simple hypothetical case study.

Objective 5. Add the Riparian Nitrogen Model as a sub-module to the SWAT_LS model for estimating nitrate removal by denitrification in riparian zones. Assess the effect of this sub-module in nitrogen simulation by running different scenarios in a hypothetical case study.

Objective 6. Apply the SWAT_LS model in the Odense river basin to evaluate if the modified SWAT model gives a better representation for flow and nitrogen simulations. Analyse the uncertainty of flow and nitrogen results for the Odense river basin using the SWAT_LS model.

1.5   OUTLINE OF THE THESIS

Chapter 1 briefly reviews research related to the field of this thesis based on which the topic of the thesis is introduced. Research questions and objectives of this thesis are listed and briefly explained. Moreover, the structure of the thesis is presented to get a brief introduction of its content.

Chapter 2 summarizes a literature review which covers several topics related to the thesis including: wetland/riparian zones and their importance; hydrological and nutrient processes happening in wetlands/riparian zones; river-basin-scale models for diffuse pollution which include detailed descriptions of the two models SWAT and DAISY-MIKE SHE that were used.

Chapter 3 presents a detailed description of the case study area of this thesis: the Odense river basin in Northern Denmark, in terms of meteorological conditions, catchment characteristics, water resources, agricultural activities and nutrient loads in the area.

Chapter 4 describes in detail the procedure to set up and calibrate the SWAT model for the Odense river basin. Flow and nitrate modelling results are presented for calibration and validation results. Moreover, a description of an existing DAISY-MIKE SHE model for the same case study is also presented in order to prepare for a comparison between the two models.

Chapter 5 compares and evaluates the simulation of flow and nitrogen fluxes in different models with different model structures. First, a comparison between SWAT which is a semi-distributed model and DAISY-MIKE SHE which is a fully-distributed physics-based model is implemented. Then, a comparison between different SWAT models with different model structures is described and evaluated.

Chapter 6 presents a modification of the SWAT model (SWAT_LS) that accounts for the landscape position of HRUs and the routing of water and nitrogen across different landscape elements. A sensitivity analysis on flow and nitrogen simulation using the SWAT_LS model in a simple hypothetical case study is implemented and compared with the original SWAT model.

Chapter 7 gives a description of a conceptual riparian zone model for simulating nitrate removal by denitrification, the Riparian Nitrogen Model (RNM), and the adding of this model into the SWAT model. Then, the performance of this modified SWAT model in a simple hypothetical case study is evaluated in different scenarios.

Chapter 8 shows an application of the modified SWAT model in the Odense river basin. A comparison of modelling results between the modified SWAT model and the original SWAT model versus measured data are also presented. An uncertainty analysis is carried out for parameters used for calibration and new parameters in the Riparian Nitrogen Model using the GLUE approach.

Chapter 9 summarises the main findings and presents conclusions and recommendations.

Chapter 2

LITERATURE REVIEW

2.1 EU WATER FRAMEWORK DIRECTIVE

The EU Water Framework Directive which was issued by EU in 2000 establishes a framework for water policy based on the principle of integrated river basin management. This Directive is an assimilation of the EU Surface Water Directive (1975), the EU Freshwater Fish Directive (1998), the EU Groundwater directive (1980), the EU Nitrate Directive (1991), the EU Urban Waste-water Treatment (1991), the EU Drinking Water Directive (1980), the new EU Drinking Water Directives (1980, 1998), and the EU Integrated Pollution Prevention and Control Directive (IPPC) (1996).

The objectives of this Directive are as follows:

✓ Expanding the scope of water protection to all waters: surface waters, coastal waters and groundwater

✓ Achieving "good status" for all waters by 2015

✓ Managing water resources at the river basin scale

✓ Combining the emission limit values approach and the quality standards approach

✓ Getting the prices right: charges for water and waste water reflecting the true costs

✓ Strengthening the participation of citizen in water management

Significant changes in this legislation are addressing pollution problems at the river basin scale and establishing water quality policies on water quality objectives (immission-based regulations) rather than on emission limit values (emission-based regulations). According to this Directive, water resources are managed according to their natural geological and hydrological unit which means the river basin scale instead of according to administrative or political boundaries, which is an effective way to include all possible sources (diffuse source and point sources) in water pollution management. Moreover, in this Directive, water resource protection changes from focusing on the control of point sources of pollution (emission-based regulations) to integrating pollution prevention at river basin level and setting water quality objectives for the receiving water (immission-based regulations).

2.2 WETLAND AND RIPARIAN ZONES

2.2.1 What is wetland?

A wetland is an ecosystem that arises when inundation by water produces soils dominated by anaerobic processes and forces the biota, particularly rooted plants to exhibit adaptations to tolerate flooding (Keddy, 2000).

This broad definition includes everything from tropical mangrove swamps to subarctic peatlands. In the definition, it can be understood that the cause of wetland is the inundation by water, a proximate effect is reduction of oxygen levels in the soil and a secondary effect is the biota must tolerate both the direct effects of flooding and the secondary effects of anaerobic conditions.

Wetlands are usually found at the interface of terrestrial ecosystems, such as upland forest and grasslands, and aquatic systems such as deep lakes and oceans, which make them different from other two ecosystems but highly dependent on both (Mitsch and Gosselink, 2000). Moreover, they are also found in seemingly isolated situations, where the nearby aquatic system is often a groundwater aquifer (figure 2.1). In all cases, the unifying principle is that wetlands are wet long enough to exclude plant species that cannot grow in saturated soils and to alter soil properties because of the chemical, physical, and biological changes that occur during flooding (Kadlec and Wallace, 2008). Figure 2.1 shows the general differences among terrestrial, wetland and aquatic system.

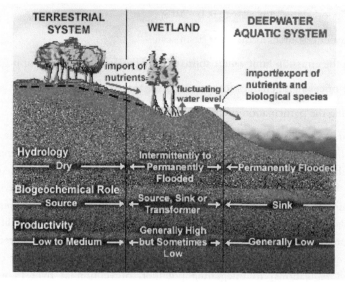

Figure 2.1 Differences among terrestrial, wetland and aquatic systems (Ramachandra et al., 2005)

The transitional position of wetlands between uplands and deepwater aquatic systems allow wetlands to function as organic exporters or inorganic nutrient sinks. Moreover, wetlands also have high biodiversity which carry characteristics of both aquatic and terrestrial systems.

Therefore, some wetlands have the distinction of being among the most productive ecosystems on Earth (Mitsch and Gosselink, 2000).

2.2.2 Wetland soil

The cause of wetlands is the inundation of water; therefore, the characteristic of the wetland soil is the hydric soil. Oxidation, aerobic decomposition, leaching and dehydration are important processes that influence the properties of soils (Keddy, 2000). All four processes are modified by flooding in wetlands, principally because water displaces air from the pore spaces between the soil particles. Because oxygen and other gases diffuse in air about 10^3 – 10^4 times faster than in water, oxygen in wetland soils is soon depleted from the flooded soil by the respiration of soil micro-organisms and plant roots. Therefore, wetland soils tend to be deficient in oxygen and form anaerobic condition. As most of the world's soil is oxidized, wetlands provide the major reducing system present in the biosphere, which gives them the function as transformers of nutrients and metals. While most terrestrial ecosystems are sources of nutrients, wetlands have the ability to store phosphorus or transform nitrogen to gases and play an important role in reducing the nutrient concentrations in the surface water systems.

2.2.3 Wetland hydrology

The water balance in a wetland can be described as follows:

$$\frac{\Delta V}{\Delta t} = P_n + S_i + G_i - ET - S_o - G_o \pm T \tag{2.1}$$

Where V = volume of water storage in wetlands

$\dfrac{\Delta V}{\Delta t}$ = change in volume of water storage in wetland per unit time, t

P_n = net precipitation

S_i = surface inflows, including flooding streams

G_i = groundwater inflows

ET = evapotranspiration

S_o = surface outflows

G_o = groundwater outflows

T = tidal inflow (+) or outflow (-)

The importance of the hydrology in wetlands

Hydrologic conditions are extremely important for the maintenance of a wetland's structure and function (Mitsch and Gosselink, 2000). The starting point of hydrology is climate and basin geomorphology. The hydrology directly modifies and determines the physiochemical environment which includes soil chemistry, water chemistry such as oxygen availability, nutrient availability, pH, toxicity etc. The hydrology also drives the transport of sediments, nutrients and even toxic materials into wetlands. Hydrology also causes water outflows from wetlands which carry biotic and abiotic material such as dissolve organic carbon, excess sediment, excess salinity, toxins. Oppositely, the physiochemical environment can also change the hydrology, for e.g. the build-up of sediments can modify the hydrology by changing basin geometry or affecting hydrologic inflows and outflows.

Changes in the physiochemical environment then have direct impact on the biota in the wetland, determining the species composition and richness and ecosystem productivity. Inversely, the biotic components of wetland can modify the physiochemistry and the hydrology. For example, wetland vegetation influences hydrological conditions by binding sediments to reduce erosion, trapping sediments, or interrupting water flows. Beavers build dams on stream and cause changes in water flow.

Generally, hydrology is an important factor in different flowing aspects of wetlands:

✓ Hydrology leads to a unique vegetation composition which is water-tolerant vegetation but can limit or enhance species richness.

✓ Primary productivity and other ecosystem functions in wetlands are often enhanced by flowing conditions and depressed by stagnant conditions.

✓ Accumulation of organic material in wetlands is controlled by hydrology through its influence on primary productivity, decomposition and export of particulate organic matter.

✓ Nutrient cycling and nutrient availability are controlled by hydrologic conditions.

Nutrients are carried into wetlands by precipitation, river flooding, tides, and surface and groundwater inflows, and out of the system by water outflows. The hydro-period which is the seasonal pattern of the water level of a wetland has significant effects on the nutrient transformation. The nitrogen availability and loss are controlled by the reduced conditions that result from waterlogged soils. Phosphorus is more soluble in anaerobic conditions due to hydrolysis and reduction of ferric and aluminium phosphates to more soluble compounds.

2.2.4 Chemical transformation in wetlands

2.2.4.1 *Oxygen and redox potential*

The inundation of soil with water results in anaerobic or reduced conditions due to the low diffusion rate of oxygen in the water. The rate of oxygen depletion depends on the ambient temperature, availability of organic substrates for microbial respiration and other chemical oxygen demands. However, oxygen is not totally depleted from the soil water of wetlands. There is usually a thin layer of oxidized soils, at the surface of the soil at the soil-water interface. The deeper layers of this layer remain reduced conditions. This oxidized layer is very important in the chemical transformations and nutrient cycling occurring in wetlands. Oxidized ions such as Fe^{3+}, Mn^{4+}, NO_3^- and SO_4^- are found in this layer while the lower anaerobic soils are dominated by reduced forms such as ferrous and manganous salts, ammonia and sulphides.

The redox potential is a quantitative measure of the tendency of the soil to oxidize or reduce substances. When organic substrates in a waterlogged soil are oxidized, the redox potential drops. The organic matter is one of the most reduced substances that can be oxidized when there is any number of terminal electron acceptors is available including O_2, NO_3^-, Mn^{2+}, Fe^{3+} or SO_4^-. Rate of organic decomposition are most rapid in the presence of oxygen and slower for electron acceptors such as nitrates and sulphates.

At a redox potential of between 400 and 600 mV, aerobic oxidation occurs for which the oxygen is the terminal electron acceptor.

$$O_2 + 4e^- + 4H^+ \rightarrow 2H_2O$$

One of the first reactions that occur in wetland soils after they become anaerobic is the reduction of NO_3^- first to NO_2^- and finally to N_2O or N_2, nitrate becomes an electron acceptor at about 250 mV:

$$2NO_3^- + 10e^- + 12H^+ \rightarrow N_2 + 6H_2O$$

When the redox potential continues to decrease below 225 mV, the transformation of manganese may occur.

$$MnO_2 + 2e^- + 4H^+ \rightarrow Mn^{2+} + 2H_2O$$

The transformation of iron from ferric to ferrous forms occurs at about +100 to -100 mV, while sulphate transformation to sulphides happens at -100 to -200 mV.

$$Fe(OH)_3 + e^- + 3H^+ \rightarrow Fe^{2+} + 3H_2O$$

$$SO_4^- + 8e^- + 9H^+ \rightarrow HS^- + 4H_2O$$

Under the most reduced conditions, the organic matter itself or carbon dioxide becomes the terminal electron acceptor at below -200 mV, producing low-molecular-weight organic compounds and methane gas.

$$CO_2 + 8e^- + 8H^+ \rightarrow CH_4 + 2H_2O$$

In addition to the redox potential, pH and temperature are also important factors that affect the rates of transformation.

2.2.4.2 pH

Organic soils in a wetland are often acidic whereas mineral soils often have neutral or alkaline condition. When a wetland is constructed, lands that were previously drained become flooded. The general consequence of flooding previously drained soils is causing alkaline soils to decrease in pH and acid soils to increase in pH and finally converging to neutral pH ranging from 6.7 to 7.2 (Mitsch and Gosselink, 2000)

2.2.4.3 Nitrogen cycle

Within a wetland, one of the principal steps controlling rates of nitrogen cycling is the rate at which organic nitrogen is mineralized to NH_4^+. Ammonification rate is much slower in flooded-soil system than in drained-soil system (Reddy, 1982). Because the depth of aerobic zone in flooded soils is usually very thin, the contribution of aerobic mineralization to the overall N mineralization is small compared to anaerobic mineralization. The rate of ammonification in wetlands is dependent on temperature, pH, C/N ratio of the residue, available nutrients, soil conditions, extracellular enzyme, microbial biomass and soil redox potential (Reddy et al., 1984). Another source for NH_4^+ is biological N_2 fixation through the activity of certain organisms in the presence of the enzyme nitrogenase. In wetland soils, fixation may occur in the floodwater, on the soil surface, in the aerobic and anaerobic flooded soils, in the root zone of plants and on the leaf and stem surfaces of plants (Buresh et al., 1980). NH_4^+ is lost though plant uptake, nitrification and volatilization to gaseous form NH_3. The two forms of nitrogen that plants can uptake are ammonia and nitrate, however, ammonia is the preferred nitrogen source as it is more reduced energetically than nitrate (Kadlec and Knight, 1996). Volatisation from NH_4^+ to NH_3 is affected by pH and temperature. An alkaline pH shifts the equilibrium towards producing more NH_3 (Middlebrooks and Pano, 1983). At a pH value of 9.5, NH_3 forms about 20% of total ammonia nitrogen at the temperature of 0°C while accounts for 70% at the temperature of 30°C (Vymazal, 2001). The nitrification processes is described in the following paragraph.

The nitrification process which is the biological oxidation of NH_4^+ to NO_3^- with NO_2^- as an intermediate happens in the shallow oxidized zone where chemoautotrophic bacteria operate. Kinetically, ammonification proceeds more rapidly than nitrification (Kadlec and Knight, 1996). Vymazal (1995) stated that the nitrification rate in wetland soils depends on the supply of NH_4^+ to the aerobic zone, pH, temperature, the presence of nitrifying bacteria, and thickness of the aerobic layer. The depletion of NH_4^+ in the aerobic layer and the larger amount of NH_4^+ from anaerobic mineralization in anaerobic layer sets up a concentration gradient which drives upward diffusion of NH_4^+ from deeper anoxic regions to the upper oxidized layer. At the same time, nitrogen in the form of NO_3^- flows in the reverse direction to the anoxic layer.

In the anoxic layer, denitrification is carried out by microorganisms with nitrate acting as a terminal electron acceptor, results in the loss of nitrogen as it is converted NO_3^- to gaseous forms N_2O and N_2 (Mitsch and Gosselink, 2000). Knowles (1982) summarized that the environmental factors that affect denitrification rates include absence of O_2, redox potential, temperature, pH, presence of nitrifiers, organic matter, nitrate concentration and inhibitors such as sulphides.

Figure 2.2 shows nitrogen cycle in flooded soil of wetlands.

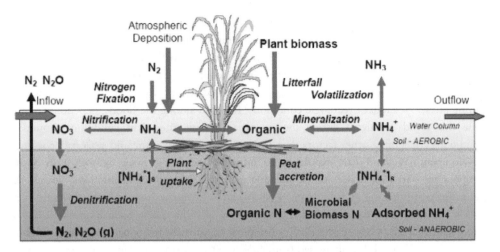

Figure 2.2 Nitrogen cycle in wetlands

Groffman et al. (1988) stated that the process of denitrification is controlled by several direct and indirect factors at different spatiotemporal scales. At process scale, the denitrification process is directly controlled by water temperature, pH and the availability of nitrate and carbon. At wetland scale, this process is controlled by the wetland position in the landscape and the wetland morphology which affects the hydraulic residence time. In order to obtain an efficient nutrient retention rate, a wetland should have a hydraulic residence time of at least 5 days (Kadlec and Knight, 1996). At landscape scale, denitrification is controlled by features of the upstream catchment area including size of the upstream basin, geological conditions, land use pattern and climate which affect the amount and temporal variability of water discharge and nutrient loads to the surface flow wetlands. Often, nitrogen removal efficiency of headwater wetland is higher than wetland located more downstream (Trepel and Palmeri, 2002). Summer months are connected to low flow and the decrease in nitrogen concentration is less pronounced at high flow occasions, which appear at wintertime (Arheimer and Wittgren, 2002). For restored wetland, the nitrogen removal capacity may be lower when the wetlands are relatively recently restored (1–10 years before monitoring) (Arheimer and Wittgren, 2002).

As mentioned above, the denitrification process can lead to the formation of N_2 and N_2O. N_2O is an intermediate product of microbial denitrification. The quantity of N_2O evolved during denitrification depends on the amount of nitrogen denitrified and the ratio of N_2 and N_2O produced. This ratio is affected by aeration, pH, temperature and nitrate to ammonia ratio in denitrifying system (Vymazal, 1995). However, N_2O is one of greenhouse gases that cause global warming. According to Van den Heuvel et al. (2009), pH is the most important regulating factor for N_2O emission from wetlands. N_2O emission decreases with the increase of pH (Knowles, 1982; Van den Heuvel et al., 2009). N_2O accounts for 80% of denitrification products including N_2 and N_2O when pH equals to 4 while it only holds less than 5% when pH reach 7 (Van den Heuvel et al., 2009).

Nitrogen retention in wetland is the result of several biogeochemical and physical processes including plant uptake, peat accumulation, denitrification and sedimentation (Howard-Williams, 1985; Mitsch and Gosselink, 2000) in which denitrification is the most significant process to improve water quality (Trepel and Palmeri, 2002). According to the study of Rücker and Schrautzer (2010), the nitrogen function of stream wetland is varied considerably during phases of similar hydrologic conditions. It varied among the floods, depending on the time that floods occurred during the season, different antecedent conditions, boundary conditions concerning other factors than hydrology, e.g. temperature and seasonal ecosystem development. Wetland can export more nitrate after the growing season. Low water temperatures and residence times meeting increase oxygen concentrations, lower the denitrification potential in the stream (Rücker and Schrautzer, 2010). Moreno-Mateos et al. (2010) evaluated the effects of wetland construction on water quality by testing in wetlands of various size (50, 200, 800, 5000m^2). Results show that nitrate retention rate were as high as 99% in the 5000m^2 while it increased from 40% to 60% for small wetlands (50, 200, 800m^2).

2.2.5 Importance of wetlands

Functions of wetlands are listed as follows:

✓ *Flood attenuation*

The hydrology is extremely important in determining the character of wetlands. Conversely, wetlands influence regional water flow regimes. Hattermann et al. (2006) stated that riparian zones and wetlands, being an interface between catchments and surface waters, can play an important role in the control of water quantity and water quality of surface water systems in general, and in the reduction of diffuse pollution in catchments in particular although they cover small parts of the total catchment. In a watershed with 5-10% of wetlands, it is possible to provide a 50% reduction in peak flood period compared to watershed without wetlands (DeLaney, 1995).

Wetland vegetation regulates stream and river flow, helping to control floods.

✓ *Erosion prevention* – Vegetation in and adjacent to wetlands and rivers slows water flow, holds soils and prevents erosion.

✓ *Aquifer recharge*

Most wetlands do not recharge groundwater systems because soils under most wetlands are impermeable. Groundwater recharge from wetlands occurs primarily around the edges of wetlands and was related to the edge: volume ratio of the wetland. Therefore, recharge is relatively more important in small wetlands than in large ones. Small wetlands can contribute significantly to recharge of regional groundwater

✓ *Improvement of water quality*

Wetland, under favourable conditions, is able to remove organic and inorganic nutrients and toxic materials from water that flow across them. At catchment scale, wetland is also a effective measures for water pollution removal for the whole catchment. Significantly better water quality exists in those watersheds where wetlands are incorporated into the landscape (Kadlec and Knight, 1996).

According to Sather and Smith (1984), several attributes that influence the chemicals that flow through wetlands and improve water quality are as follows:

1. A reduction in water velocity as streams enter wetlands, causing sediments and chemicals sorbed to sediments to drop out of water column

2. A variety of anaerobic and aerobic processes which include denitrification, chemical precipitation, and other chemical reactions that remove chemicals from the water

3. A high rate of productivity that can lead to high rates of mineral uptake by vegetation and subsequent burial in sediments when the plants die

4. A diversity of decomposers and decomposition processes in wetland sediments

5. A large contact surface of water with sediments because of the shallow water, leading to significant sediment - water exchange

6. An accumulation of organic peat in many wetlands that causes the permanent burial of chemicals

As discussed above, hydrology is an important factor for water quality processes in wetlands. Moreover, the efficiency of wetlands in water quality improvement depends on temperature as increasing temperature accelerates all metabolic processes, including denitrification and biological uptake. According to the study of Arheimer and Wittgren (2002), due to temperature dependence and seasonal variation in water discharge, significant decrease in nitrogen concentrations mainly occurred during summer periods with low loading. This is an interesting issue to study in this research to see how the wetland performances change in different season with different weather conditions and different input loadings to the wetland.

✓ *Climatic stability*

Wetland vegetation can act as a carbon reservoir and assists in reducing the amount of carbon dioxide in the atmosphere, decreasing the greenhouse effect and leading to a more stable climate.

✓ *Linear oases*

Both perennial and ephemeral rivers that pass through otherwise arid areas are sources of water and support linear strips of vegetation, enabling people and wildlife to survive there.

This study will focus mostly on the functions of wetlands to improve water quality and control water quantity at the catchment scale.

2.2.6 Riparian zones

Riparian zone is the interface between terrestrial and aquatic ecosystems representing an important ecological component of the landscape. Although not easily delineated, the riparian zone encompasses the vegetated strip of land that extends along streams and rivers (Gregory et al., 1991). This location, between uplands and aquatic ecosystems, allows riparian zones the capability of modifying effects on the aquatic environment. The importance of the riparian zone exceeds the minor proportion of the land base it occupies (Imhol et al., 1996; Roth et al., 1996).

2.2.6.1 Influences of riparian zones

One of the ways the riparian zones influences an adjacent waterway is through its vegetation. Forested riparian zones are important regulators of the microclimate of adjacent streams (Imhol et al., 1996; Roth et al., 1996). By intercepting solar radiation, trees in the riparian zone modify the magnitude of solar input and significantly lower stream temperature (Gregory et al., 1991). Since several in-stream processes are directly related to stream temperature, the riparian zone has an indirect impact on other aspects of the aquatic environment. Vegetation is also an important source of organic and inorganic materials through particulate terrestrial inputs. The removal of riparian vegetation eliminates an important segment of food for aquatic invertebrates (Roth et al., 1996).

Riparian zones have the ability to trap sediments amassed in upslope areas. As runoff from urban and agricultural activities passes through a vegetated riparian strip, the velocity of overland flow is greatly reduced causing sediment to fall out of suspension (Osborne and Kovacic, 1993). Moreover, vegetation in the riparian zone can prevent erosion at stream side by putting down roots that stabilise the stream bank (Vought et al., 1994).

Because of their physical location between uplands and aquatic systems, riparian zones are in an ideal position to modify, incorporate, dilute or concentrate substances in groundwater before they enter waterways. Where surface runoff is prevalent, riparian zones are able to trap sediment and thus prevent sediment-bound pollutants from reaching a stream. The "living filter" property of riparian zones is vital not only to the stream adjacent the riparian zone, but to downstream receiving water bodies as well. Therefore, the planting and/or preservation of riparian buffers strips has been recommended as an effective means of reducing pollution from agricultural land (Osborne and Kovacic, 1993; Vought et al., 1994).

More widely studied is the effect of riparian buffer zones on the discharge of nutrients, particularly nitrate into freshwater system (Haycock and Burt, 1993; Osborne and Kovacic, 1993). Nitrate can originate from geological sources, precipitation, mineralization of organic nitrogen and agricultural activities. Nitrate is a highly water soluble anion repelled by the net negative charge in most soils and therefore, is highly mobile. Nitrate is also able to persist in groundwater for decades and can accumulate to high levels if more nitrogen is applied to the land surface each year. Numerous studies have examined the effect of riparian zones on nitrogen content of subsurface water flow in natural catchments (Haycock and Burt, 1993; Jordan et al., 1993; Osborne and Kovacic, 1993). These studies show that buffer zones are consistently able to reduce NO_3 to below 2 mg/l. Low concentrations of nitrate have been reported in riparian-zone groundwater not only in undisturbed headwater watersheds (Campbell et al., 2000; McDowell et al., 1992; Sueker et al., 2001) but also in agricultural watersheds (Hill, 1996; Jordan et al., 1993). Three possible mechanisms explaining the retention of nitrate in riparian zones are: plant uptake, microbial immobilisation and bacteria denitrification.

Hill (1996) indicated nitrogen retention by vegetation does not necessarily indicate uptake of N from groundwater as other N inputs could be responsible for biomass accumulation. Osborne and Kovacic (1993) compared grass and forested riparian zones and found that both were effective filters for NO_3 although the grass filter strips are less effective at filtering nitrate from shallow groundwater than the forested one on annual basis. On the other hand, Groffman et al. (1991) observed that denitrification rates were consistently higher in grass plots than forest pots. This finding is supported by Schnabel et al. (1997) who also found that grassy sites had greater denitrification rates than the woody site.

Of the three mechanisms for nitrate attenuation, microbial immobilisation may be of minor importance (Hill, 1996; Jordan et al., 1993). Occurring under anaerobic conditions, immobilisation is a dissimilatory reduction of NO_3 to NH_4 that can reduce nitrate concentration in groundwater (Hill, 1996). Jordan et al. (1993) did not observe a corresponding in NH_4 as NO_3 decreased at their riparian site, implying that the immobilisation process does not have significant effect which is also agreed by Groffman et al. (1996).

Among the three mechanisms, the most commonly cited mechanism of nitrate removal is bacteria denitrification (Groffman et al., 1991; Vought et al., 1994). While microbial immobilisation and uptake by vegetation likely have a supporting role in the retention of NO_3, biological denitrification is the main mechanism for groundwater nitrate attenuation.

2.2.6.2 Denitrification in riparian zones

Because the end product of the denitrification process is N_2, a gas that is abundant in the earth's atmosphere, denitrification is considered the most desirable means of nitrate removal from drainage waters. The direct and indirect factors of denitrification process were mentioned in the nitrogen cycles of wetlands in section 2.2.4.3.

✓ *Temporal variability of denitrification in riparian zones*

The temporal variability of riparian zone denitrification is an important consideration. It has been suggested that in temperate climates, the most concentrated nitrate discharges occur in the winter months when the bioassimilation of nitrogen by plants is not possible (Groffman et al., 1992). Burt and Arkell (1987) reported that 80% of NO_3 leached from agricultural soil was concentrated during winter. Pinay et al. (1993) and Groffman et al. (1992) also found peak denitrification times during the winter months in the surface soils of the southwest of France and Rhode Island, respectively. However, several studies reported different seasonal pattern for denitrification. Groffman et al. (1996) observed denitrification to be significantly higher in June and September than in March and February at Kingston, Rhode Island, USA. Another study in Kingston, Rhode Island observed inconsistent seasonal patterns over a two year period with lowest rates of denitrification occurring in the summer months (June to

August) at a very poorly drained site and lowest rates of denitrification occurring in the autumn (September to November) at a poorly drained site (Groffman and Hanson, 1997).

✓ *Spatial variability of denitrification in riparian zones*

Parkin (1987) and Gold et al. (1998) demonstrated that 'hot-spots' of denitrification activity are associated with patches of organic carbon in the soil. The variability in denitrification with depth tend to focus on the surface sediments (Hill, 1996; Pinay et al., 1993). However, measurements made only surfacemost soil may deviate from the pathways of groundwater flow and nutrient discharge (Hill, 1996). Geyer et al. (1992) reported that large masses of agricultural NO_3 can be transported to depth of 16 m below cultivated fields via groundwater percolation. Therefore, it is clear that subsurface riparian soils must be considered when examining the overall effectiveness of riparian zone on NO_3 removal. The potential of the attenuation of nitrate in groundwater through riparian zones cannot be based solely on the denitrification capacity of surface soils (Martin et al., 1999).

2.2.6.3 Modelling of denitrification in riparian zones

Nitrogen inputs have been increasing all over the world due to the demand of food and energy production activities supporting the growing population (Galloway et al., 2004). The changing nitrogen cycle and associated abundance of reactive nitrogen in the environment has been linked to many concerns including the deterioration of air quality (Townsend et al., 2003), disruption of forest ecosystem processes (Aber et al., 2003), acidification of lakes and streams (Driscoll et al., 2001) and degradation of coastal waters including eutrophication, hypoxia and harmful algal blooms (Rabalais, 2002). Therefore, denitrification is a very important process because it is the only mechanism by which reactive forms of N in terrestrial and aquatic landscapes are transformed back to N_2 which is the dominant component of the earth's atmosphere.

Quantifying where, when and how much denitrification occurs in ecosystems is particularly difficult in all spatial scales (Groffman et al., 2006; Seitzinger et al., 2006). It is difficult to detect changes in N_2 in the environment attributed to denitrification. Presently, there is no scientific method for making direct measurements of the rates of denitrification at large scales. Even at field scales, there remain challenges in using direct measurement of denitrification. Because the controlling factors of denitrification are highly variable over space and time, they give rise to "hot spots" and "hot moments" of denitrification that are difficult to predict (McClain et al., 2003)

Terrestrial and aquatic models have become essential tools for integrating current understanding of the processes that control denitrification with broad scale measurements of the rate-controlling with properties so that the losses of N can be quantified within landscape (Boyer et al., 2006). Models provide a framework for extrapolating over a wide range of spatial and temporal scales, and over a range of climatic, soil and land use conditions. In this thesis, only denitrification in the terrestrial ecosystems is concerned. Boyer et al. (2006) identified a variety of terrestrial landscape models to illustrate the range of approaches that have been used to quantify the rates of N flux and denitrification in soils and terrestrial ecosystems from field to regional spatial scales as below:

✓ *Mass balance model*

Based on the inputs, outputs, and changes in storage in landscapes, mass balance budget is used to explore the magnitude of denitrification occurring over large areas. This method was used in variety of studies such as Howarth et al. (1996), Boyer et al. (2002) and van Breemen et al. (2002) to calculate N losses within the soils and groundwater of the terrestrial landscape as the difference between N inputs and all storage and loss terms. Van Breemen et al. (2002) indicated that denitrification was the most likely mechanism for explaining N losses in the terrestrial landscape.

✓ *DAYCENT model*

The DAYCENT model represents the full N cycle and both short and long term time scales of soil organic matter dynamics. The denitrification sub-module used in DAYCENT, original call NGAS, was first presented by Parton et al. (1996). The denitrification sub-module assumes that N gas flux from denitrification is controlled by soil NO_3 concentration (e^- acceptor), labile C availability (e^- donor) and O_2 availability (competing e^- acceptor). The model first calculates total N gas flux from denitrification, then partitions this between N_2 and N_2O using an $N_2:N_2O$ ratio function. The ratio function assumes that as O_2 availability decreases, a larger proportion of N_2O from denitrification will be further reduced to N_2 before diffusing from the soil to the atmosphere.

The DAYCENT model has been tested to simulate N cycling at field sites and over regional scales (Del Grosso et al., 2002; Del Grosso et al., 2005). The results showed that the denitrification sub-module satisfactorily simulated N_2 emissions for the date used for model parameterisation. The ability of DAYCENT to simulate N_2 needs to be further tested, but field data describing N_2 fluxes over space and time are very limited.

✓ *DNDC model*

The denitrification-decomposition (DNDC) model was initially developed for quantifying nitrous oxide (N_2O) emissions from agriculture soils in the United States (Li et al., 2000; Li et al., 1992; Li et al., 1996). The core of DNDC is a soil biogeochemistry model, containing two components to bridge between primary drivers and the coupled biogeochemical cycles of carbon and nitrogen in terrestrial ecosystems. The first component, consisting of soil climate, plant growth, and decomposition sub-models, predicts the soil environmental factors using the primary drivers as input parameters. The second component consists of nitrification, denitrification and fermentation sub-models, quantifies production and consumption of N_2O, NO, N_2, ammonia and methane by tracking the kinetics of relevant biochemical or geochemical reactions, driven by the modelled soil environmental factors.

Denitrification is modelled with a series of biologically mediated reduction reaction from nitrate to N_2. The key equations adopted in DNDC for modelling the microbial activities include (i) the Nernst equation which is a basic thermodynamic formula defining soil Eh based on concentrations of the oxidants and reductants existing in a soil liquid phase; and (ii) Michaelis-Menten equation describing the kinetics of microbial growth with dual nutrients.

✓ *EPIC and related model*

The EPIC model (Sharpley and Williams, 1990; Williams et al., 1984) is a dynamic simulation model that describes the influence of agricultural management on crop productivity and erosion. EPIC simulates the major N cycling processes in agricultural soils including mineralisation, nitrification, immobilisation, ammonia volatilisation and denitrification, runoff and subsurface leaching.

In EPIC, denitrification is simulated as a function of nitrate availability, C availability, soil temperature, and soil moisture content. If the ratio of soil water content to field capacity in a soil layer is greater than 95% or soil water content is greater than 90% of the saturation value indicating nearly saturation conditions and likely anoxia, denitrification can occur.

The field scale agricultural management model GLEAMS (Leonard et al., 1987) was developed from both EPIC and CREAMS and employs a more explicit description of soil water content. In GLEAMS, denitrification only occurs if the soil water content is greater than a parameter related to the soil water content at field capacity and saturation. The EPIC and GLEAMS method of simulating denitrification neglects denitrification that may occur in anaerobic micro-zones when the soil is not at field capacity or saturation. Therefore, EPIC and GLEAMS will tend to estimate a lower frequency of denitrification than observations, but overestimate the magnitude of denitrification when soil water content exceeds the threshold defined for denitrification.

The SWAT model incorporates features of CREAMS, GLEAMS, EPIC to describe land surface and subsurface processes and uses components of the QUAL2E model to simulate in-stream and reservoir transport of contaminants. The N loss from denitrification is estimated for individual soil layers as a function of the initial nitrate concentration in the soil water, temperature, and organic C percentage. Denitrification losses increase with increase in temperature and C.

The DRAINMOD model (Skaggs, 1999) and the new version of the model, DRAINMOD-N II (Youssef et al., 2005) quantify N losses and transport from agricultural lands with shallow water tables where artificial drainage systems are extensively used. DRAINMOD-N II considers both NO_3-N and NH_x-N in modelling mineral N and simulates nitrification and denitrification processes. Denitrification is modelled using Michaelis-Menten kinetics for NO_3-N. The influence of organic C on the process rate is implicitly expressed in the exponential soil depth function. The maximum denitrification rate is site specific and depends on the soil organic matter content and texture and agronomic practices.

✓ *INCA and RHESSys models*

The INCA (integrated nitrogen in Catchments) model which is a water and N mass balance simulation model estimates the integrated effects of point and diffuse N sources on stream nitrate and ammonium concentrations and loads and also estimates the N loads related to processes in the plant/soil system (Whitehead et al., 1998). Denitrification is modelled according to a first order function of soil wetness and the nitrate concentration of the soil water; the denitrification rate coefficient is a mass flux expressed as length per time.

The regional hydrological ecosystem simulation system (RHESSys) (Band et al., 1991) is also widely used to explore N dynamics at the watershed scale. Streamflow generation and the flowpath partitioning of overland flow, through flow, and baseflow is based on the implementation of variable source area concepts based on topography, quantifying routing of water through the landscape from patch to patch. Band et al. (2001) applied RHESSys in a forested watershed in Maryland to simulate denitrification rates over time and space. Similar to the INCA model, RHESSys also lacks explicit representation of groundwater volumes and residence time, and thus does not quantify N that is denitrified in groundwater.

✓ *The Riparian Nitrogen model*

The Riparian Nitrogen Model (RNM) (Rassam et al., 2008) is a conceptual model that estimates the removal of nitrate as a result of denitrification. The denitrification occurs when groundwater and surface waters interact with riparian buffers. This interaction occurs via two mechanisms: (i) groundwater passes through the riparian buffer before discharging to the stream and (ii) surface water is temporarily stored within the riparian soils during flood event. This model will be added as a sub-module of the SWAT model in this thesis. Its detailed description can be seen in chapter 7.

2.3   RIVER BASIN SCALE MODELS FOR DIFFUSE POLLUTION

River basin scale models which are capable of estimating pollutant loads from diffuse sources in the basin to the river are necessary in sustainable environmental management. Yang and Wang (2010) reviewed several well known, operational and free modeling tools that are able to handle diffuse water pollution. The CREAMS model (Knisel, 1980) was developed for the analysis of agricultural best management practices (BMP) for pollution control. GLEAMS (Leonard et al., 1987) which was built based on the CREAMS model can be considered as the vadose zone component of the CREAMS model. The CREAMS and GLEAMS models are limited to the scale of a small field plot. SWMM (Huber and Dickinson, 1988) is an urban model that can simulate quantity and quality processes in the urban hydrologic cycle. The ANSWERS (Beasley and Huggins, 1981) model is capable of predicting the hydrologic and erosion response of agricultural river basins. The AGNPS (Young et al., 1986) model can handle both point sources and diffuse sources and can be used to estimate nutrients and sediments in runoff and compare the effects of various pollution control practices in river basin management. The SWRRB model was developed by modifying CREAMS for evaluating basin scale water quality by daily simulation of hydrology, crop growth, nitrogen, phosphorus and pesticide movement (Williams et al., 1985). The Soil and Water Assessment Tool (SWAT) developed by the USDA Agricultural Research Service (ARS) is a physically-based model to simulate the impact of land management activities on water quantity and quality, sediment transport, pesticides and nutrient leaching in large complex river basins over long time period (Arnold et al., 1998). A lot of studies have been carried out to use SWAT to calculate nutrient loads (Huang et al., 2009; Salvetti et al., 2008; Yang et al., 2009) and suggest measures to improve water quality by running SWAT models with different management scenarios (Ullrich and Volk, 2009; Volk et al., 2009). SHETRAN is a 3D physical based, spatially distributed model to simulate water flow, sediment transport and solute transport in river basins (Ewen, 2000). A nitrogen transformation model, NITS (Nitrate Integrated Transformation component for SHETRAN) added to SHETRAN by Birkinshaw and Ewen (2000) is capable to simulate concentration of nitrogen species along with water flow and nitrogen transport. The HSPF model (Bicknell et al., 2000) is one of the most detailed, operational models of agricultural runoff, erosion and water quality simulation. The DAISY-MIKE SHE (Styczen and Storm, 1993a, 1993b) approach is a sequentially coupled model of a physically-based root zone model DAISY (Abrahamsen and Hansen, 2000; Hansen et al., 1991) and a physically-based and fully distributed river basin model MIKE SHE (Refsgaard and Storm, 1995). In addition to SHETRAN, the coupled DAISY-MIKE SHE model is also a physically based distributed model with 3D groundwater module and is able to model the nitrogen transport and removal by denitrification in aquifers. Conan et al.(2003) developed an integrated model which consists of SWAT for water and nitrogen fate in the unsaturated zone, MODFLOW as groundwater flow using time-variant recharge predicted by SWAT; and MT3DMS for simulating the nitrate fate leached from the topsoil as predicted by SWAT.

Two river basin models with different approaches were used in this thesis: SWAT and DAISY-MIKE SHE. The main case study of this thesis, Odense river basin in Denmark, is occupied by agriculture areas with more than 50% of the area drained by tile. Therefore, the selected models need to be able to handle tile drainage simulation. The DAISY-MIKE SHE model was already set up and simulation results are available for the Odense river basin. In the following sections, the two models SWAT and DAISY-MIKE SHE are described in details.

2.3.1 SWAT (Soil and Water Assessment Tool)

The soil and water assessment tool (SWAT) is a physically based, time continuous model, developed by the USDA Agricultural Research Service (ARS) in order to simulate the impact of land management activities on water, sediment, pesticides and nutrient yields in large complex watersheds over long time period. In SWAT, a watershed is divided into multiple sub-basins. They are then subdivided into hydrological response units (HRUs) each of which have unique land cover, soil characteristic and management combination. All processes modeled in SWAT are lumped at HRU level (Neitsch et al., 2004).

In SWAT, the hydrological cycle is the driving force behind whatever happens in the watershed. Simulation of the hydrology of a watershed can be separated into two major divisions. The first division is the land phase of the hydrologic cycle which controls the amount of water, sediment, nutrient and pesticide loadings to the main channel in each sub-basin. The second one is the water or routing phase of the hydrologic cycle which can be defined as the movement of water, sediments, etc. through the channel network of the watershed to the outlet (Neitsch et al., 2004). The transformation processes of water quality components are modelled in the routing phase with QUAL2E model concept.

SWAT model is a very useful tool to calculate the pollution loads from diffuse sources. A lot of studies have been carried out to use SWAT to calculate nutrient loads and suggest measures to improve water quality by running SWAT models with different management scenarios. Huang et al. (2009) got reasonable results for stream flow and nutrient loadings, Yang et al. (2009) also obtained good results in simulated water flow and sediment yield as well as variation of soluble phosphorus. However, the simulated nitrogen and water soluble phosphorus is generally higher than measured value due to the wetland processes in riparian zones (Yang et al., 2009). SWAT is able to represent general trend of water quality changes result from different management scenarios, thus evaluate the effect of management practices alternative on the watershed level (Ullrich and Volk, 2009; Volk et al., 2009). Kang et al. (2005) applied SWAT for TMDL programs to a small watershed containing rice paddy fields. In this study, the SWAT model was used to calculate nutrient loads which originated from the animals and application of manure and fertilizer in the rice paddy fields. The result was used to compare with Total Maximum Daily Loads and the requirement for decreasing nutrient loads was proposed in each sub-basin. Bouraoui and Grizzetti (2008) used SWAT to identify the major processes and pathways controlling nutrient losses from agriculture activities. Salvetti et al. (2008) also used SWAT as a tool to the rain-driven diffuse load (load from runoff and erosion processes).

2.3.1.1 Land phase of the hydrological cycle

The land phase of hydrological cycle in SWAT simulates the loading of water, sediment, nutrients and pesticides from each sub-basin to the main channel.

✓ **Water balance**

The hydrological cycle is based upon the water balance (equation 2.2 and figure 2.3).

$$SW_t = SW_0 + \sum_i^t \left(R_i - Q_{surf,i} - E_{a,i} - w_{seep,i} - Q_{gw,i} \right) \tag{2.2}$$

Where SW_t is the final soil water content (mm), SW_0 is the initial water content on day i (mm), t is the time (days), R_i is the amount of precipitation on day i (mm), $Q_{surf,i}$ is the amount of surface runoff on day i (mm), $E_{a,i}$ is the amount of evapotranspiration on day i (mm), $w_{seep,i}$ is the amount of percolation on day i (mm) and $Q_{gw,i}$ is the amount of base flow on day i (mm).

As precipitation descends, it is intercepted by plant cover and then evaporated or fall to the soil surface. Water falling to the soil surface will infiltrate into the soil profile or form surface runoff which is considered as a relatively quick flow to stream channel. Infiltrated water will be uptaken by plant and then evapotranspired, evaporate through soil holes, form lateral flow in subsurface layer or continue to percolate to the shallow aquifer. Water reaching shallow aquifer can move back to the soil profile through capillary rise and uptake of deep-rooted plants or form groundwater flow that contributes to the stream or continue to infiltrate to the deep (confined) aquifer.

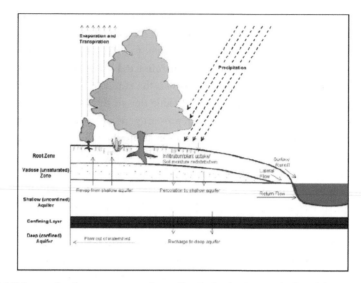

Figure 2.3 Schematization representation of hydrological cycle in SWAT (Neitsch et al., 2004)

In daily time step, SWAT simulates surface runoff using the SCS curve number method (USDA-NRCS, 2004) in which canopy storage is taken into account in the surface runoff calculation. SWAT assigns the SCS curve number based on land use, hydrologic soil group and hydrologic condition. The amount of infiltration to the soil profile is the difference between the amount of rainfall and surface runoff. The percolation component of SWAT uses a storage routing technique to predict flow through each soil layer in the root zone. Percolation occurs when the field capacity of a soil layer is exceeded and the layer below is not saturated. The flow rate is governed by the saturated conductivity of the soil layer. Lateral subsurface flow in the soil profile is calculated simultaneously with redistribution using a kinematic storage model. The model computes evapotranspiration separately for soil and plants. Potential soil water evaporation is estimated as a function of potential evapotranspiration and leaf area index. Actual soil water evaporation is estimated by using exponential functions of soil depth and water content. Plant transpiration is simulated as a linear function of potential evapotranspiration and leaf area index. For groundwater, SWAT partitions groundwater into two aquifers: a shallow, unconfined aquifer which contributes return flow to stream within the watershed and a deep, confined aquifer which contributes return flow to streams outside the watershed. Water percolating past the bottom of the rootzone is partitioned into two fractions each of which becomes recharge for one of the aquifers. Moreover, water in shallow aquifer can be directly removed by plants or move to overlying unsaturated layer when water stored in shallow aquifer exceed a threshold value. Water in shallow or deep aquifer can be removed by pumping (Neitsch et al., 2004).

✓ **Nitrogen balance**

SWAT monitors five different pools of nitrogen in the soil (figure 2.4). Two inorganic forms of nitrogen are NH_4 and NO_3 and 3 organic forms of nitrogen are fresh organic N which is associated with crop residue and microbial biomass, active and stable organic N associated with the soil humus.

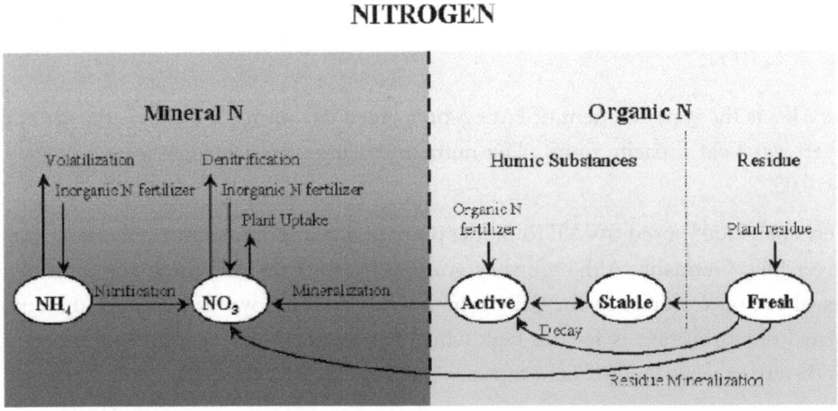

NITROGEN

Figure 2.4 SWAT nitrogen pools and nitrogen processes in land phase (Neitsch et al., 2004)

Nitrogen processes and transport are modelled by SWAT in the soil profile, in the shallow aquifer and in the river reaches. Nitrogen processes simulated in the soil include mineralization, residue decomposition, immobilization, nitrification, ammonia volatilization and denitrification. Ammonium is assumed to be easily adsorbed by soil particles and thus it is not considered in the nutrient transport. Nitrate, which is very susceptible to leaching, can be lost through surface runoff, lateral flow and percolate out of the soil profile and enter the shallow aquifer. Nitrate in the shallow aquifer may also be lost due to uptake by the presence of bacteria, by chemical transformation driven by the change in redox potential of the aquifer and other processes. These processes are lumped together to represent the loss of nitrate in the aquifer by the nitrate half-life parameter.

✓ **Denitrification simulated in SWAT**

SWAT determines the amount of nitrate lost to denitrification with the equation:

$$N_{denit,ly} = NO3_{ly} \cdot \left(1 - \exp\left[-\beta_{denit} \cdot \gamma_{tmp,ly} \cdot orgC_{ly}\right]\right) \quad if \ \gamma_{sw,ly} \geq \gamma_{sw,thr} \tag{2.3}$$

$$N_{denit,ly} = 0.0 \quad\quad\quad\quad\quad\quad if \ \gamma_{sw,ly} < \gamma_{sw,thr} \tag{2.4}$$

where $N_{denit,ly}$ is the amount of nitrogen lost to denitrification (kg N/ha), $NO3_{ly}$ is the amount of nitrate in layer ly (kg N/ha), β_{denit} is the rate coefficient for denitrifcation, $\gamma_{tmp,ly}$ is the nutrient cycling temperature factor for layer ly, $\gamma_{sw,ly}$ is the nutrient cycling water factor for layer ly, $orgC_{ly}$ is the amount of organic carbon in the layer (%) and $\gamma_{sw,thr}$ is the threshold value of nutrient cycling water factor for denitrification to occur.

$\gamma_{tmp,ly}$ and $\gamma_{sw,ly}$ are calculated as belows:

$$\gamma_{tmp,ly} = 0.9 \cdot \frac{T_{soil,ly}}{T_{soil,ly} + \exp\left[9.93 - 0.312 \cdot T_{soil,ly}\right]} + 0.1 \qquad (2.5)$$

where $T_{soil,ly}$ is the temperature of layer ly (^{o}C). The nutrient cycling temperature factor is never allowed to fall below 0.1.

$$\gamma_{sw,ly} = \frac{SW_{ly}}{FC_{ly}} \qquad (2.6)$$

where SW_{ly} is the water content of layer ly on a given day (mm), and FC_{ly} is the water content of layer ly at field capacity (mm). The nutrient cycling water factor is never allowed to fall below 0.05.

Pohlert et al. (2005) used SWAT to model point and non-point source pollution of nitrate in the river Dill, Germany. Although they got good results on daily flow and monthly nitrate fluxes, the model efficiency for daily nitrate loads is very low. In addition, the amount of simulated denitrification is far too high when it is assumed that denitrification occurs when soil moisture exceeds 95% of field capacity ($\gamma_{sw,ly} = 0.95$).

Ferrant et al. (2011) compared nitrogen dynamics in a small agricultural catchment between using a distributed model TNT2 and semi-distributed model SWAT. They illustrated the uncertainties of using ago-hydrological models to simulate catchment water chemistry and the equifinality problem, i.e. different model structures can reproduce outlet flows and nitrate loads with different internal dynamics (mineralization and denitrification dynamics in this study) although simulated annual water and N yields are very close. The results confirmed that the use of such tools for prediction must be considered with care, unless a proper calibration and validation of the major N processes is carried out.

2.3.1.2 Routing phase of the hydrological cycle

✓ **Water routing**

After SWAT determines the loadings of water, sediments, nutrients and pesticides to the main channel, the loadings are routed through the stream network of the watershed. In addition to keeping track of mass flow in the channel, SWAT also models the transformation of chemicals in the stream and streambed.

As water flows downstream, a portion may be lost due to evaporation and transmission through the bed of the channel. Another potential loss is removal of water from the channel for agricultural or human use. Flow may be supplemented by the fall of rain directly on the channel and/or addition of water from point source discharges.

Routing in the main channel is divided into 4 components: water, sediment, nutrients and organic chemicals/pesticides. However, this study only considered water routing and nitrogen routing.

Originally, flow is routed through the canal using the *variable storage routing method* or the *Muskingum routing system*. These methods are variations of the kinematic wave model. Therefore SWAT cannot model backwater effect.

Variable Storage method

In the Variable Storage method, outflow depends on the stored volume in the reach and the inflow and a calculated Storage Coefficient (SC).

$$V_{out,2} = SC(V_{in} + V_{stored,1}) \tag{2.7}$$

where SC is the storage coefficient that depends on the travel time TT and the time step Δt, V_{in} is the average inflow at the beginning and the end of the time step (m³), $V_{out,2}$ is the outflow at the end of the time step (m³), $V_{stored,1}$ is the storage volume at the beginning of the time step (m³).

$$SC = \frac{2 \cdot \Delta t}{2 \cdot TT + \Delta t} \tag{2.8}$$

$$TT = \frac{V_{stored}}{q_{out}} = \frac{V_{stored,1}}{q_{out,1}} = \frac{V_{stored,2}}{q_{out,2}} \tag{2.9}$$

Where TT is the travel time (s), V_{stored} is the storage volume (m³) and q_{out} is the discharge rate (m³/s)

Muskingum method

In the Muskingum method, the outflow is calculated from the inflow, the inflow of previous time step and the outflow of previous time step. Storage in the reach is thus not a required state for this method.

$$q_{out,2} = C1 \cdot q_{in,2} + C2 \cdot q_{in,1} + C3 \cdot q_{out,1} \tag{2.10}$$

where $q_{in,1}$ is the inflow rate at the beginning of the time step (m³/s), $q_{in,2}$ is the inflow rate at the end of the time step (m³/s), $q_{out,1}$ is the outflow rate at the beginning of the time step (m³/s), $q_{out,2}$ is the outflow rate at the end of the time step (m³/s). C1, C2 and C3 depend on the time step Δt and K, the storage time constant for the reach segment (s) and X, a weighting factor that controls the relative importance of inflow and outflow in determining the storage in a reach.

$$C1 = \frac{\Delta t - 2 \cdot K \cdot X}{2 \cdot K \cdot (1 - X) + \Delta t} \tag{2.11}$$

$$C2 = \frac{\Delta t + 2 \cdot K \cdot X}{2 \cdot K \cdot (1 - X) + \Delta t} \tag{2.12}$$

$$C3 = \frac{2 \cdot K \cdot (1 - X) - \Delta t}{2 \cdot K \cdot (1 - X) + \Delta t} \tag{2.13}$$

where $C1 + C2 + C3 = 1$

To maintain numerical stability and avoid the computation of negative outflows, the following condition must be met:

$$2 \cdot K \cdot X < \Delta t < 2 \cdot K \cdot (1 - X) \tag{2.14}$$

X is input by the user. The value of K is estimated as:

$$K = coef_1 \cdot K_{bnkfull} + coef_2 \cdot K_{0.1bnkfull} \tag{2.15}$$

Where $coef_1$ and $coef_2$ are weighting coefficients input by the user, $K_{bnkfull}$ is the storage time constant calculated for the reach segment with bankfull flows (s), $K_{0.1bnkfull}$ is the storage time constant calculated for the reach segment with one-tenth of the bankfull flows (s). $K_{bnkfull}$ and $K_{0.1bnkfull}$ are calculated using the following equation:

$$K = \frac{1000 \cdot L_{ch}}{c_k} \tag{2.16}$$

where L_{ch} is the channel length (km), c_k is the celerity corresponding to the flow for a specified depth (m/s) which is calculated as follows:

$$c_k = \frac{5}{3} \cdot \left(\frac{R_{ch}^{2/3} \cdot slp_{ch}^{1/2}}{n} \right) = \frac{5}{3} \cdot v_c \tag{2.17}$$

where R_{ch} is the hydraulic radius for a given depth of flow (m), slp_{ch} is the slope along the channel length (m/m), n is Manning coefficient for the channel, and v_c is the flow velocity (m/s)

✓ Nitrogen routing

SWAT model tracks nutrients dissolved in the stream and nutrients absorbed to the sediment. Soluble nutrients are transported with the water while nutrient adsorbed to sediments are allowed to be deposited with the sediment on the streambed.

Nutrient transformation in the stream is controlled by in-stream water quality components of the model. The in-stream kinetics used in SWAT for nutrient routing is adapted from QUAL2E model (Brown and T.O. Barnwell, 1987).

QUAL2E model is based on the Streeter-Phelps equations. It is a steady-state, 1-dimensional model to simulate flow and water quality in streams and rivers that can be assumed to be well-mixed. The variables and processes in QUAL2E are divided in 3 layers: phenomenological models with variables BOD and DO; biochemical level with ammonia, nitrite, nitrate and Sediment Oxygen Demand; ecological level about algae model. The limitation of QUAL2E is that the mass balances are not always consistent because of different levels. Moreover, the use of BOD as organic carbon variable is not suitable for calculating mass balances because BOD only refers to organic matter that can be biologically decomposed (Van Griensven and Bauwens, 2003). The nitrogen cycles modelled in SWAT are described as follows.

The nitrogen cycle in QUAL2E model is composed of 4 components: organic nitrogen, ammonia nitrogen (NH_4^+), nitrite nitrogen (NO_2^-) and nitrate nitrogen (NO_3^-) (Figure 2.5)

The amount of organic nitrogen in the stream is increased by the conversion of algal biomass nitrogen to organic nitrogen. Organic nitrogen is then mineralized to NH_4^+ or settles with sediment. Ammonium (NH_4^+) is then converted to NO_2^- and from NO_2^- to NO_3^- through nitrification. Moreover, there may be diffusion of ammonium from streambed sediments which increases the amount of ammonium. The conversion of nitrite to nitrate occurs more rapidly than the conversion of ammonium to nitrite, so the amount of nitrite present in the stream is usually very small. Ammonium and nitrate are uptaken by algae as necessary nutrients for their lives.

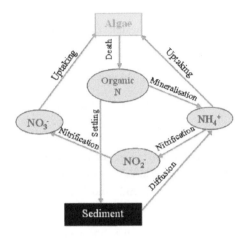

Figure 2.5 In-stream nitrogen cycle in SWAT

2.3.1.3 *Wetland/ riparian zone modelling in SWAT*

✓ **Wetland modelling in SWAT**

SWAT models wetlands and ponds as water bodies located within sub-basins that received inflow from a fraction of the sub-basin area.

The water balance for a wetland is

$$V = V_{stored} + V_{flowin} - V_{flowout} + V_{pcp} - V_{evap} - V_{seep} \qquad (2.18)$$

where V is volume of water in the wetland at the end of the day (m³), V_{stored} is the volume of water stored in the water body at the beginning of the day (m³), V_{flowin} is the volume of water entering the water body during the day (m³), $V_{flowout}$ is the volume of water flowing out of the water body during the day (m³), V_{pcp} is the volume of precipitation falling on the water body during the day (m³), V_{evap} is the volume of water removed from the water body by evaporation during the day (m³), and V_{seep} is the volume of water lost from the water body by seepage (m³).

The water entering a wetland on a given day is calculated:

$$V_{flowin} = fr_{imp} \cdot 10 \cdot (Q_{surf} + Q_{gw} + Q_{lat}) \cdot (Area - SA) \qquad (2.19)$$

where V_{flowin} is the volume of water entering a wetland on a given day (m³), fr_{imp} is the fraction of sub-basin area draining into the impoundment, Q_{surf} is the surface runoff from the sub-basin on a given day (mm), Q_{gw} is the groundwater generated in the sub-basin on a given day (mm), Q_{lat} is the lateral flow generated in the sub-basin on a given day (mm), $Area$ is the sub-basin area (ha) and SA is the surface area of wetland (ha). The volume of water entering the wetland is subtracted from the surface runoff, lateral flow and groundwater loadings to the main channel.

The wetland releases water whenever the water volume exceeds the normal storage volume, V_{nor}. Wetland outflow is calculated:

$$
\begin{aligned}
V_{flowout} &= 0 & &\text{if } V < V_{nor} \\
V_{flowout} &= \frac{V - V_{nor}}{10} & &\text{if } V_{nor} \leq V \leq V_{mx} \\
V_{flowout} &= V - V_{mx} & &\text{if } V > V_{mx}
\end{aligned}
\qquad (2.20)
$$

where $V_{flowout}$ is the volume of water flowing out of the wetland during the day (m³), V is the volume of water stored in the wetland (m³), V_{mx} is the volume of water held in the wetland when filled to the maximum water level (m³), and V_{nor} is the volume of water held in the wetland when filled to the normal water level (m³).

In order to calculate nutrient transformation in the wetland, SWAT assumes the system is completely mixed. In the completely mixed system, as nutrients enter the wetland, they are instantaneously distributed throughout the volume. Nutrient transformations simulated in wetlands in SWAT are limited to the removal of nutrients by settling. Transformation between nutrient pools including organic nitrogen, nitrate, nitrite and ammonia are ignored.

✓ **Current riparian zone modelling in SWAT**

Filter strips module

The version of SWAT2005 (Neitsch et al., 2005) gives an option to model filter strips. In the filter strips module, sediment, nutrient and pesticide retentions are modelled using similar efficiency calculated from the below equation

$$trap_{ef} = 0.367 \cdot (width_{filterstrip})^{0.2967} \qquad (2.21)$$

where $trap_{ef}$ is efficiency of sediment, nutrient and pesticide retentions; and $width_{filterstrip}$ is width of filter strips (m).

Filter strips can also reduce loads of constituents in subsurface flow that pass through the strip. The trapping efficiency for subsurface flow constituents ($trap_{ef,sub}$) is calculated:

$$trap_{ef,sub} = \frac{(2.1661 \cdot width_{filterstrip} - 5.1302)}{100} \qquad (2.22)$$

where $trap_{ef,sub}$ is efficiency of sediment, nutrient and pesticide retentions for subsurface flow; and $width_{filterstrip}$ is width of filter strips (m).

SWAT_VFS

White and Arnold (2009) developed a field scale Vegetative Filter Strips (VFS) sub-model for SWAT. In this sub-model, the model for the retention of sediments and nutrients in VFSs was developed from experimental observations derived from 22 publications. The runoff retention model was developed from Vegetative Filter Strip MOdel (VFSMOD) simulations.

- ***Runoff reduction model***

Due to this lack of data, the runoff reduction model was derived from the database of VFSMOD simulations.

$$R_R = 75.8 - 10.8\ln(R_L) + 25.9\ln(K_{sat}) \qquad (2.23)$$

where R_R is the runoff reduction; R_L is the runoff loading (mm); and K_{sat} is the saturated hydraulic conductivity (mm/h).

- ***Sediment reduction model***

The sediment reduction model developed for SWAT was based on measured VFS data. Sixty-two experiments reported in the literature were used to develop this model.

$$S_R = 79.0 - 1.04S_L + 0.213R_R \qquad (2.24)$$

where S_R is the predicted sediment reduction (%); S_L is sediment loading (kg/m^2); and R_R is the runoff reduction (%).

- ***Nitrogen reduction models***

The nitrate nitrogen model was developed from 42 observations. The current version of VFSMOD does not account for nutrients.

$$NNR = 39.4 + 0.584R_R \qquad (2.25)$$

where NNR is the nitrate nitrogen reduction (%); and R_R is the runoff reduction (%).

From equation 2.25, there is a minimum reduction of 39.4% in nitrate, even if there is no reduction in runoff due to the VFS. This outcome may be unexpected, but it is supported by the measured data (White and Arnold, 2009).

2.3.2 DAISY- MIKE SHE

DAISY- MIKE SHE (Styczen and Storm, 1993b) approach is a coupling model of a physically-based root zone model DAISY (Abrahamsen and Hansen, 2000; Hansen et al., 1991) and a physically-based and fully distributed catchment model MIKE SHE (Refsgaard and Storm, 1995). The coupling between two models was implemented in order to simulate groundwater water quality within a hydrological river basin.

2.3.2.1 DAISY model

DAISY is a 1-dimensional agro-ecosystem model that, simulates crop production and crop yield, and describes water dynamics, soil temperature, the carbon and nitrogen cycle of the root zone in agricultural soil based on information on management practices and weather data. The schematic overview DAISY model system can be seen in figure 2.6.

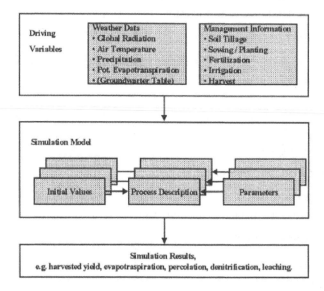

Figure 2.6 The schematic overview DAISY model system (Abrahamsen and Hansen, 2000)

✓ **Water balance**

The adopted schematization of the water dynamics in agricultural soils is illustrated in figure 2.7. The boxes represent storage of water, and the arrows represent flow of water or water vapour.

The hydrological processes simulated in DAISY include snow accumulation and melting, interception of precipitation by the crop canopy, evaporation from crops and soil surface, infiltration, water uptake by plant roots, transpiration and vertical movement of water in the soil profile.

Precipitation and irrigation represent driving variables or boundary conditions. Another driving variable or boundary condition is the potential evapotranspiration which forms both the driving force and the upper limit in the evaporation/transpiration processes. In the evapotranspiration calculations, evaporation from free water surfaces has priority over transpiration and soil evaporation.

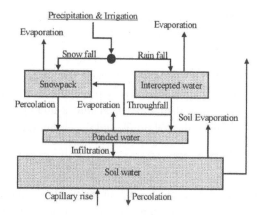

Figure 2.7 Schematic representation of the hydrological cycle in DAISY model (Abrahamsen and Hansen, 2000)

✓ **Nitrogen cycle**

The transformation and transport processes considered in DAISY include net mineralisation of nitrogen, nitrification, denitrification, nitrogen uptake by plants and nitrogen leaching from the root zone.

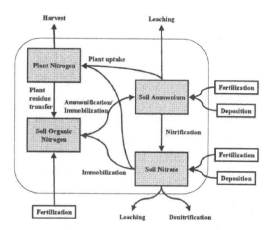

Figure 2.8 Schematic representation of Nitrogen cycle components included in DAISY (Abrahamsen and Hansen, 2000)

2.3.2.2 MIKE SHE

MIKE SHE is a physically based and fully distributed hydrological catchment model. It simulates the flow of water and solutes in the catchment by solving the governing equations of overland and channel flow, unsaturated and saturated flow by finite difference methods. The model can perform numerical solutions for 3D Boussinessq equation for saturated flow, 1D Richard equation for unsaturated zone, 2D Saint Venant equation for overland flow and is integrated with MIKE 11 to model the exchange flow and transport between the river and the saturated zone. The overview of hydrological processes modeled in MIKE SHE can be seen in figure 2.9.

The variation in catchment characteristics (e.g. land use, soil, geology, topography) and driving variables (e.g. climatic input data) are represented in a network of grids in the horizontal direction. Each grid is then subdivided into a number of layers in the vertical direction to describe variations in the soil profile and the groundwater aquifer system.

✓ **Modelling in land phase**

MIKE SHE allows to simulate all of the processes in the land phase of the hydrologic cycle. Precipitation falls as rain or snow depending on air temperature - snow accumulates until the temperature increases to the melting point, whereas rain immediately enters the dynamic hydrologic cycle. Initially, rainfall is either intercepted by leaves (canopy storage) or falls through to the ground surface. Once at the ground surface, the water can now either evaporate, infiltrate or runoff as overland flow. If it evaporates, the water leaves the system. However, if it infiltrates then it will enter the unsaturated zone, where it will be either extracted by the plant roots and transpired, added to the unsaturated storage, or flow downwards to the water table. If the upper layer of the unsaturated zone is saturated, then additional water cannot infiltrate and overland flow will be formed. This overland flow will follow the topography downhill until it reaches an area where it can infiltrate or until it reaches a stream where it will join the other surface water. Groundwater will also add to the baseflow in the streams, or the flow in the stream can infiltrate back into the groundwater.

✓ **Modelling in main channels**

In order to model the main channels or rivers in the catchment, MIKE SHE is fully coupled with MIKE 11 in which data were exchanged between the two models after each computational time step. MIKE 11 is a one-dimensional river model designed for simulation of flow, transport and water quality. It is based on the complete dynamic wave formulation of the Saint Venant equations. The coupling model between MIKE SHE and MIKE 11 has the ability to model wetlands and is a useful tool for wetland planning for water quality improvement.

For water quality modelling, MIKE 11 has an Ecolab submodule for ecological simulations. Ecolab enables the simulation of various in-stream processes such as denitrification, nitrification, and degradation of organic matter. Furthermore, it is possible to apply user-defined processes such as denitrification in wetlands adjacent to the river system.

Thompson et al. (2004) applied this integrated model for a lowland wet grassland, the Elmley Marshes, in southeast England. The model performed well in hydrodynamic modelling. However, little is known about modelling water quality in wetlands using this model.

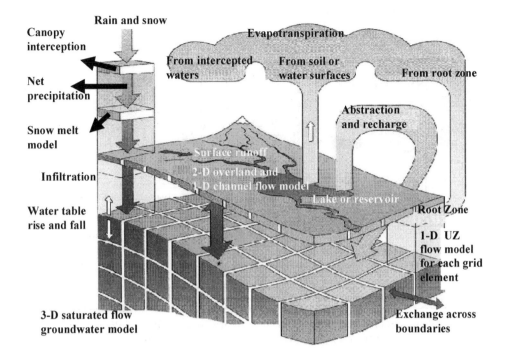

Canopy
interception

Rain and snow

Evapotranspiration

From intercepted
waters

From soil or
water surfaces

From root zone

Net
precipitation

Abstraction
and recharge

Snow melt
model

Surface runoff
2-D overland and
1-D channel flow model

Infiltration

Lake or reservoir

Water table
rise and fall

Root Zone

1-D UZ
flow model
for each grid
element

3-D saturated flow
groundwater model

Exchange across
boundaries

Figure 2.9 MIKE-SHE modelling framework (Refsgaard and Storm, 1995)

2.3.2.3 Coupling of MIKE-SHE and DAISY

MIKE-SHE is coupled with the crop model DAISY to simulate the crop yield and nitrogen cycle in the basin in the integrated DAISY-MIKE SHE modelling approach. The two models are coupled sequentially without any feedback from the groundwater and river to the root zone, as described in figure 2.10. Groundwater levels are used as boundary conditions in DAISY, which are also simulated by MIKE SHE in the saturated zone. The pipe drainage (tile drainage) option can be applied in DAISY as one of the upper boundary conditions. However, this does not result in simulated tile drainage flow in the coupled DAISY-MIKE SHE model; instead, it regulates infiltration of water from the root zone in DAISY to the vadose zone in MIKE-SHE. Tile drainage flow is simulated in DAISY-MIKE SHE when the groundwater level rises above the tile drain depth in MIKE SHE. The DAISY model first simulates the water and nitrogen budget of the root zone for all the specified combinations of input data. The DAISY column simulations are then distributed to corresponding geo-referenced field blocks within the basin. The outputs from field blocks are then aggregated as daily net values to MIKE SHE model grid blocks using a weighting area procedure. Subsequently, flow and transport simulation are executed in MIKE SHE.

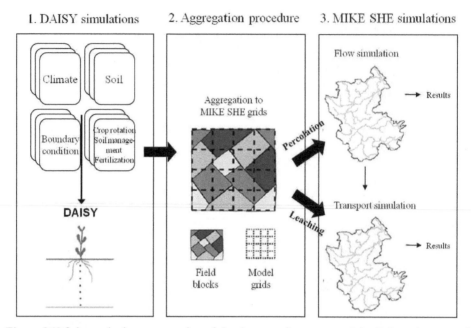

Figure 2.10 Schematical representation of the data requirements and the linkage between the DAISY and MIKE-SHE (Styczen and Storm, 1993a)

2.4 INTEGRATED WETLANDS AND RIPARIAN ZONES IN RIVER BASIN MODELLING

In order to evaluate the importance of wetlands and riparian zones in the reduction of diffuse pollution in the catchment and the improvement of the water quality in the surface water, the integration of wetland and riparian zones in river basin modelling is essential.

Hattermann et al. (2006) implemented wetlands and riparian zones in SWIM (Soil and Water Integrated Model) that is a modified version of SWAT (Soil and Water Assessment Tool) by extending SWIM to reproduce the relevant water and nutrient flows and retention processes catchment scale and in riparian zones and wetlands. The main extensions introduced in SWIM were: (1) implementation of daily groundwater table dynamics at the hydrotope level, (2) implementation of water and nutrient uptake by plants from groundwater in riparian zones and wetlands, and (3) assessment of nutrient retention in groundwater and interflow.

Arheimer and Wittgren (2002) assessed the nitrogen removal in potential wetlands at the catchment scale by incorporating wetland models into a dynamic process-based catchment model (HBV-N). The wetland model applied was very simple in which wetland was treated as a batch reactor and N removals was assumed to be area dependent. The study also applied wetland models modified from the basic model by including nitrogen re-suspension and Arrhenius temperature dependence; however, the modifications gave no substantial support.

An integrated and physically based nitrogen cycle catchment model which is based on a combination of physically based root-zone model DAISY (Hansen et al., 1991) and a physically based distributed catchment model MIKE SHE (Refsgaard and Storm, 1995) was established and evaluated as a potential approach for optimal planning of the establishment of wetlands and further land use management with respect to high denitrification rates (Hansen et al., 2009). In addition to simulate denitrification in the saturated zone, this approach can also simulate denitrification processes in wetlands and sediments in the river bed.

Wetland modelling is also included in SWAT which models wetlands as a part of the sub-basin covering it. However, it is limited to remove nutrients by settling and ignores the biochemical processes. SWAT also gives an option to model filter strips using filter strip module (Neitsch et al., 2005) which model nutrient and subsurface flow trapping efficiencies using empirical equations, and a more complicated model SWAT- VFS in which empirical equations for runoff, sediment and nutrient reduction are developed from experimental observations derived from 22 publications.

Rassam et al. (2005) introduced Riparian Nitrogen model (RNM) which is a conceptual model that estimates the removal of nitrate as a result of denitrification, which is one of the major processes that lead to the permanent removal of nitrate from shallow groundwater during interaction with riparian soils. Rassam et al. (2008) linked RNM with the E2 Modelling framework (Argent et al., 2005) to evaluate the role of riparian zones on river basin water quality. E2 is a node-link model with ability to predict the hydrologic behaviour of river basins. Sub-basin processes are modelled by a combination of up to three types of processes: runoff generation, contaminant generation and filtering. The former two components produce daily time series for discharge and contaminant load in the basin which are modelled by E2. The filtering component is modelled by RNM as a "plug-in" filter option for E2 to evaluate the role of riparian buffers on nitrate loads in streams.

Chapter 3

STUDY AREA: ODENSE RIVER BASIN, DENMARK

Figure 3.1 The Northern part of Odense river basin viewed from the south towards Odense city. Odense Fjord can be seen at the top of the photograph. Photograph: Jan Kotod Winther

The Odense River basin is located on the island of Funen, the second largest Danish island. The basin comprises an area of approx. 1,050 km² corresponding to one third of Funen with about 246,000 inhabitants. The town Odense is the largest population centre in the basin area. There are about 1,015 km of watercourse in the river basin, including the largest river on Funen, the river Odense, which is about 60 km long and drains a catchment of 625 km².

Figure 3.1 shows northern part of the Odense river basin viewed from the south part and figure 3.2 presents the location of the Odense River basin.

3.1 DESCRIPTION OF ODENSE RIVER BASIN

3.1.1 Climate

The average monthly precipitation in Odense River Basin varies between approximately 40 mm (April) and 90 mm (December/ January). A large part of the precipitation evaporates, especially in summer, and only a minor share reaches the watercourses. As a consequence, the variation in monthly riverine runoff is considerably greater than the variation in precipitation. In summer the riverine runoff is therefore typically only around 20% of that in the winter months.

The average air temperature in Funen is 8.2°C (1961—1990). The wind usually blows from the west.

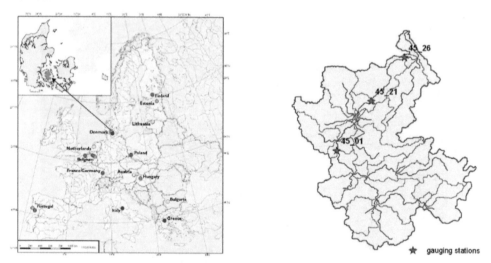

Figure 3.2 The Odense River Basin (Fyns county, 2003)

3.1.2 Soil type

The present landscape of Fyn was primarily created during the last glacial period 11,500 to 100,000 years ago. The most common landscape feature is moraine plains covered by moraine clay that was deposited by the base of the ice during its advance. The meltwater that flowed away from the ice formed meltwater valleys. An example is the Odense floodplain, which was formed by a meltwater river that had largely the same overall course as today's river. The clay soil types are slightly dominant and encompass approximately 51% of the basin, while the sandy soil types cover approximately 49% (figure 3.3). The moraine soils of Fyn are particularly well suited to the cultivation of agricultural crops. Agriculture has therefore left clear traces in the landscape. Deep ploughing, liming and the suchlike have thus rendered the surface soils more homogeneous.

Coarse sandy and
fine sandy soil

Clayey sand soil

Sandy clay soil

Clay soil

Special soil type

Humic soil

Unclassified:

Urban areas

Woodland

Inland waters (lakes,
watercourses

Residual areas

Figure 3.3 Soil types in Odense River Basin (Fyns county, 2003)

3.1.3 Land use

Land use in the Odense River Basin is dominated by agricultural exploitation. Farmland thus accounts for 68% of the basin. Of the remainder, approximately 16% is accounted for by urban areas, 10% by woodland, and 6% by natural/ semi-natural areas (meadows, mires, dry grasslands, lakes and wetlands, which are protected by Section 3 of the Protection of Nature Act). The corresponding figures for farmland, woodland and natural/semi-natural areas for Denmark as a whole are 62%, 11% and just over 9%, respectively. Figure 3.4 illustrates the distribution of land use types in the Odense river basin.

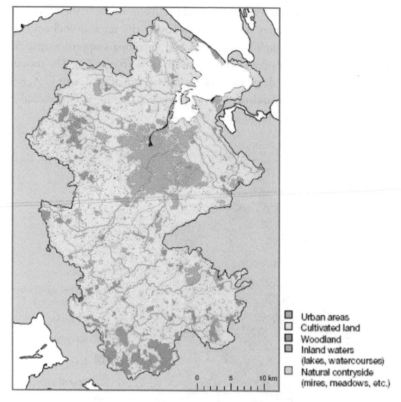

Urban areas
Cultivated land
Woodland
Inland waters
(lakes, watercourses)
Natural contryside
(mires, meadows, etc.)

Figure 3.4 Land use in Odense river basin (Environment Centre Odense, 2007)

3.1.4 Population

The population of the Odense River Basin numbers approximately 246,000, of which about 182,000 inhabit the city of Odense, which is Denmark's third largest city. The distribution of urban areas and their population density are shown in figure 3.5. Ninety percent of the population in the river basin discharges their wastewater to a municipal wastewater treatment plant. The remaining 10% of the population live outside the towns in areas not serviced by the sewerage system. A total of approx. 6,900 residential buildings are located in these sparsely built-up areas outside the sewerage system catchments.

Due to the increasing industrialization and the spread of water-flushed toilets at the beginning of the 20th Century, the amount of wastewater discharged into the water bodies of Funen from towns, dairies, abattoirs, etc. increased markedly. In the 1950s, moreover, agriculture really started to pollute the aquatic environment through the discharge of silage juice, slurry and seepage water from manure heaps. Later the many dairies and abattoirs were closed down through centralization, and serious efforts were initiated to treat urban wastewater. The main progress came in the 1980s and early 1990s, which saw marked improvement in the treatment of urban and industrial wastewater and the cessation of unlawful agricultural discharges of silage juice, etc.

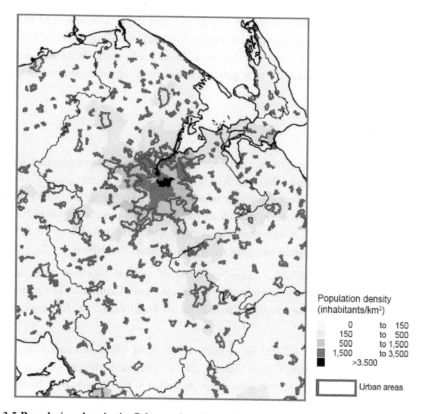

Figure 3.5 Population density in Odense river basin (Environment Centre Odense, 2007)

3.1.5 Artificial drainage and land reclamation in the river basin

Artificial drainage is estimated to have been established on at least 55% of the arable land in the Odense river basin over the past 50-100 years to ensure rapid drainage of the arable land and optimize the possibilities to cultivate it. Moreover, mires, meadows, watercourses, shallow lakes and fjord sections have undergone considerable physical modification or have completely disappeared due to land reclamation for agricultural purposes. This has resulted in the disappearance of 72% of the former large meadows and mires in the Odense river basin over the past 100 years. A large proportion of the former meadows/mires in the river valley have been converted to arable land through watercourse regulation and regular water. Therefore, long reaches of the River Odense are highly physically modified. Around 70% of the natural watercourses in Funen have been regulated or culverted. At least 25% of the watercourses are culverted. Of the remaining open watercourses, 60% are estimated to be regulated (straightened, deepened, etc) (Environment Centre Odense, 2007). The number of small lakes and ponds has also decreased considerably. Since the end of 19th century, 13 large lakes in Odense River Basin have been drained. The area of water surface in Odense Fjord has decreased by about 30% since the 1770s, and the former fjord bed has been converted to farmland through dyking and drainage.

In conclusion, land reclamation, drainage of wetlands and the establishment of field drainage during the past decades have considerably reduced the self-cleansing ability of Odense Fjord and the river basin. However, the re-establishment of wetlands and restoration of watercourse nowadays is expected to improve nutrient retention and turnover in the coming years.

3.1.6 Water courses

The watercourses in the river basin are typical lowland water courses (defined by a terrain elevation of less than 200m). The total length of watercourse is just over 1,000 km. The largest watercourse is the River Odense (drained area 625 km²) which is about 60 km long and up to 30 m wide. Watercourse density in the river basin is approximately 1.0 km/km². The natural density of the watercourse network probably used to be up to 50% greater, however, watercourse regulation and culverting have reduced the density.

There are a total of 2,620 lakes larger than 100m², which together cover an area of approximately 11 km², corresponding to 1% of the total river basin area. The majority of the lakes are small, only 21 of the lake in the river basin are larger than 3 ha (Environment Centre Odense, 2007).

Odense river basin contains 2,203 ha of mire, 1,743 ha of freshwater meadow and 481 ha coastal meadow (Environment Centre Odense, 2007). Compared with the country as a whole, the wetland habitat types are relatively weakly represented in the Odense river basin.

Studies performed by Fyn County show that the area of mire, freshwater meadow and coastal meadow has decreased by approximately 70% since the 1940s and currently accounts or approximately 5% of the river basin. As a result, the large contiguous area of natural countryside has become much smaller and lie isolated from each other separated in particular by arable land. Small, isolated areas of natural countryside are unable to maintain the same flora and fauna as large areas. Therefore, since the end of the 19th century, numerous plant species associated with wetland habitats have died out on Funen.

Thirty-six groundwater bodies have been identified within the Odense river basin. These groundwater bodies together cover an area of 722 km², corresponding to 69% of the total river basin area. Nearly all the groundwater bodies are located in sand aquifers. The majority of the groundwater bodies (29) are in contact with surface waters, typically with lakes and watercourses. Therefore, the quality and quantity of the groundwater can influence the quality and environmental status of the surface water bodies. Among the 29 groundwater bodies, 26 are in contact all year round, and the remaining are only in contact during the winter period.

3.2 PRESSURES ON WATER QUALITY

The pressures on water bodies in the Odense River Basin include input of pollutants (organic matters, nutrients and hazardous substances) and physical pressures on the water bodies (land reclamation, drainage, watercourse maintenance, water abstraction, shipping, etc). Pollutant loading to watercourses origin from both diffuse sources (e.g. agriculture, atmospheric deposition and groundwater recharge) and point sources (e.g. wastewater discharges from households and industries, effluents from wastewater treatment plants, leaching from landfills etc). Relative to total point-source and diffuse loading of Odense river basin, point sources account for a considerably greater proportion in the summer half-year than during the winter half-year. According to nitrogen, point sources account for an average of 20% of the total load in the summer and 10% in the winter while the corresponding figures for phosphorus are about 45% and 25%, respectively.

The primary sources of pollutants are described in details as follows.

3.2.1 Wastewater from households and industry

Wastewater pressure on the water bodies derives from wastewater treatment plants, stormwater outfalls from separate and combined sewerage systems and from sparsely built-up areas and industries. Wastewater contains pollutants including organic matter, nitrogen, phosphorus, hazardous substances, heavy metals and pathogenic bacteria and viruses. Since the end of the 1980s, the total amount of BOD_5, nitrogen and phosphorus in wastewater discharged into Odense River Basin has decreased considerably due to the improved treatment at wastewater treatment plants.

There are 25 wastewater treatment plants larger than 30 person equivalent (PE) in Odense River Basin. 10 of them are smaller than 100 PE while 8 are larger than 10,000 PE in which Ejby Molle is the largest with the capacity of 325,000 PE. This plant treats nearly 75% of all the wastewater discharged into the public sewerage system in the river basin. The magnitude of treated wastewater discharges depends on the amount and intensity of precipitation during the year.

There are 489 registered stormwater outfalls in which 204 are from combined sewerage systems and 285 are from separate sewerage system. The discharges vary from year to year depending on the precipitation. Approximately 6,900 properties are located in sparsely built-up areas of Odense River Basin.

Moreover, there is also a contribution of wastewater leaching from the Stige Ø Landfill. Previously, the landfill was established in 1965 without a bed membrane and therefore, nitrogen, hazardous substances and heavy metals possibly leached from the landfill to Odense Fjord/ Odense Canal. However, a system for draining the landfill was completed in 2006.

Table 3.1 shows the discharges of pollutants from point sources in Odense river basin in the year 2002.

Table 3.1 Point sources loading of the surface water bodies in Odense river basin (Environment Centre Odense, 2007)

Source	BOD$_5$ (tons/yr)	Nitrogen (tons/yr)	Phosphorus (tons/yr)
Wastewater treatment plants	63	137	6
Stormwater outfalls	116	40	10
Sparesely built-up areas	141	36	8
Industry (Stige Ø Landfill)	26	164	2
Total	**346**	**377**	**26**

3.2.2 Agriculture

In 2000, there were approx. 1,870 registered farm holdings in the Odense River Basin, of which approx. 960 were livestock holdings. The livestock herd in the river basin numbered approx. 60,000 livestock units (1999–2002), of which 59% was accounted for by pigs, 37% by cattle and 4% by other livestock. Livestock density averages 0.9 livestock unit per hectare farmland, corresponding to the national average. However, livestock density in the individual sub-catchments of the Odense River Basin varies.

Overall, the livestock production in the river basin has increased in the recent years. However, this increase marks a decrease in livestock production in the cattle sector and a marked increase in pig production. Based on the applications submitted to the authorities for expansion of livestock herds and the sector's own expectations, livestock production is expected to increase further in the coming years. The predominant crop in the river basin is cereals (approx. 2/3 winter cereals), encompassing 63% of the arable land, while 10% is permanent grassland. The concentration of market gardens is relatively high in Odense River Basin, accounting for approximately 3% of the arable land.

Agricultural activities are the dominant source of nitrogen pressure on terrestrial natural habitats and the aquatic environment. Thus, agriculture accounts for approximately 70% of the total water-borne nitrogen loading of surface waters in the river basin (2003 - 2004) and half or more of atmospheric deposition of nitrogen on water bodies and terrestrial natural habitats. Regarding to phosphorus loading of the water bodies, agricultural activities account for approximately 25% of all waterborne phosphorus loading of the water bodies (2000 - 2004) (Environment Centre Odense, 2007).

3.2.3 Atmospheric deposition

Atmospheric pollutants are deposited in the form of either wet or dry deposition. The pollutants emitted to the atmosphere from sources such as industry, power stations, households, traffic and agriculture will eventually be deposited on the land or surface water. The pollutants deposited from atmosphere can derive from local sources or transport from other areas beyond the boundary of the river basin. For example, ammonia emitted from agriculture is largely deposited locally whereas nitrogen oxides originating from power stations and traffic are largely transported afar.

The total atmospheric deposition of nitrogen (N) and phosphorus (P) on the Funen landmass is approximately calculated to be at average value of 20 kg N/ha/yr and 0.2 kg P/ha/yr from 2000-2003 (Environment Centre Odense, 2007). Due to the difference in roughness, the deposition of pollutant is generally higher on terrestrial natual habitats and woodland than on farmland, and the deposition on water surface is less than land surface.

3.3   TOTAL NUTRIENT LOADS

Annual riverine nitrogen and phosphorus loading of Odense Fjord from diffuse sources and point sources is shown in figure 3.6. Diffuse loading varies considerably from year to year, primarily due to interannual variation in precipitation and runoff. The mean freshwater runoff to the fjord from 1976 – 2005 is approximately 305 mm/yr. The mean annual precipitation in the river basin is 825 mm.

Riverine phosphorus loading of the fjord has decreased by approximately 80% since the beginning of 1980s while nitrogen loading has decreased by 30 – 35% (Environment Centre Odense, 2007). The reduction in phosphorus loading is due to the fact that the wastewater is now treated far more effectively than previously. The reduction in nitrogen loading is the combined result of improved wastewater treatment and reduced leaching from arable land due to the implementation of the Action Plans on Aquatic Environment.

Figure 3.7 shows the percentage of nitrogen and phosphorus loading from different sources. According to the nitrogen source apportionment, agriculture is the main source of diffuse loading, accounting for about 70% of the total load. Point sources account for approximately 13%, and the natural background load accounts for about 18%.

Different from nitrogen, over 30% of the riverine phosphorus load is accounted for by natural background loading. Agriculture accounts for approximately 25% while the remaining 45% is accounted for wastewater discharges from sparsely built-up areas, stormwater outfalls and municipal wastewater treatment plants.

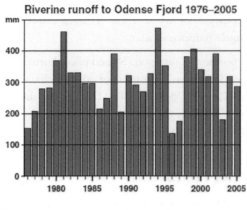

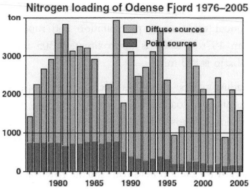

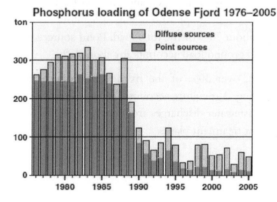

Figure 3.6 Freshwater runoff, riverine nitrogen and phosphorus loading of Odense Fjord in the period 1976-2005 apportioned between diffuse and point sources (Environment Centre Odense, 2007)

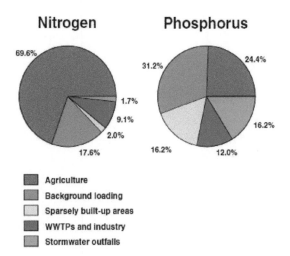

Nitrogen

69.6%
1.7%
9.1%
2.0%
17.6%

Phosphorus

31.2%
24.4%
16.2%
16.2%
12.0%

- Agriculture
- Background loading
- Sparsely built-up areas
- WWTPs and industry
- Stormwater outfalls

Figure 3.7 Source apportionment of riverine nitrogen and phosphorus loading of Odense Fjord from 1999/2000 – 2003/2004

Chapter 4

MODEL SET-UPS FOR THE ODENSE RIVER BASIN

4.1 SWAT MODEL SET-UP FOR THE ODENSE RIVER BASIN

The basic procedure to set up the SWAT model includes three main steps:

- River basin delineation: in this step, the river basin is divided in multiple sub-basins, each of which has one outlet and one point source.

- HRU definition: sub-basins are subsequently divided into HRUs based on soil type, land use type and slope. All the processes in SWAT are calculated at HRU level.

- Locating climate stations: in this step, precipitation, temperature, solar radiation, wind speed and relative humidity stations available in the area are located. Then SWAT chooses the station that is used as input to each sub-basin based on the distance between each sub-basin and climate stations. A sub-basin uses the data from the nearest climate station.

Moreover, additional inputs relating to the specific case study such as point sources, fertilizer application, reservoir operation, wetlands, etc are also added to model setups. The detailed procedure for setting up the SWAT model for the Odense river basin is described below.

4.1.1 SWAT model setup procedure

River basin delineation

The ArcSWAT interface (Olivera et al., 2006) was used to delineate the Odense river basin based on an automatic procedure using Digital Elevation Model (DEM) data. A DEM grid map with a spatial resolution of 100 meters was available. Moreover, a mask map which identifies the focused area for delineation to reduce the processing time and a burn-in river map which helps to accurately predict the location of the stream network were also input. Eventually, the Odense river basin was divided into 30 sub-basins, in each of which one point source was added. Three out of 30 outlets of sub-basins were manually added at the locations of three gauging stations (45_01, 45_21 and 45_26) which were then used for model calibration (figure 4.1). Data for nine point sources which represented discharges from sewerage systems in the basin were added in nine sub-basins, the remaining point sources were assumed to not release any flow or pollutants.

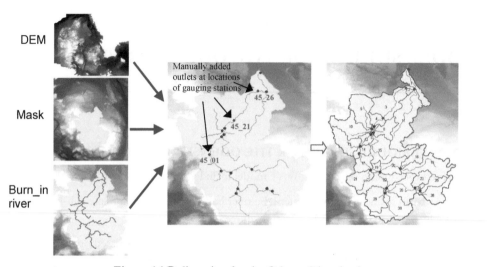

DEM

Mask

Burn_in
river

Manually added
outlets at locations
of gauging stations

45_26

45_21

45_01

Figure 4.1 Delineation for the Odense River basin

HRU definition

The Hydrological Response Units (HRUs) in SWAT are defined on the basis of soil type, land use and slope classifications.

Soil: The available data for the soil profile in the Odense river basin was divided into three horizons: A (0-30 cm), B (30-70 cm) and C (70-150 cm). The soil in each horizon was classified into different soil classes based on the percentage of clay, silt and sand according to the Danish soil classification (Greve and Breuning-Madsen, 2005) which is shown in table 4.1. In order to decrease the number of soil profiles in the model and the number of HRUs created, it was assumed that the distribution of soil types in horizon B and C is the same as in horizon A. Soil characteristics and hydraulic parameters of each soil type in each horizon were estimated by averaging all the values of the same soil type in the same horizon. As a result, there were seven soil types distributed in the SWAT model: JB1 coarse sand (0.9%), JB3 coarse clayey sand (31.6%), JB4 fine clayey sand (20.5%), JB5 coarse sandy clay (26.6%), JB6 fine sandy clay (19.4%), JB7 clay (1%), and JB11 organic (0.03%).

Land use: The land use map was taken from the existing DAISY-MIKE SHE model which divided the area into 7 types of land use: cattle farms, plant production, pig farms, grass, coniferous forest, deciduous forest and urban area. In each of the four agricultural groups, one crop rotation was simulated (table 4.2). The set-up was different from the DAISY-MIKE SHE model in which the crop rotation schemes were permuted to represent each crop every year. The detail of crop permutation can be found in the DAISY set-up description (Hansen et al., 2007) and is also mentioned in the below section. However, with the difficulty of including different crop rotations in each land use in one SWAT model set-up, this issue was not considered in this study. 50% of the urban areas were assumed to be covered by pavement or buildings which release 100% of surface runoff to the streams while the remaining 50% is assumed to be covered by permanent grass.

Table 4.1 Danish soil classification (Greve and Breuning-Madsen, 2005)

Map color code	Soil type	JB no.	Percentage by weight				
			Clay < 2 µm	Silt 2-20 µm	Fine sand 20-200 µm	Total sand 20-2000 µm	Humus 58.7%C
1	Coarse sand	1	0-5	0-20	0-50	75-100	
2	Fine sand	2			50-100		
3	Clayey sand	3	5-10	0-25	0-40	65-95	
		4			40-95		
4	Sandy clay	5	10-15	0-30	0-40	55-90	≤ 10
		6			40-90		
5	Clay	7	15-25	0-35		40-85	
6	Heavy clay or silt	8	25-45	0-45		10-75	
		9	45-100	0-50		0-55	
		10	0-50	20-100		0-80	
7	Organic soil	11					> 10
8	Atypic soil	12					

Table 4.2 Types of land use and their crop rotations in Odense river basin

No.	Type of land use	Percentage of the basin	Crop rotation
1	Cattle farms	11.3	Spring Barley (year 1), Grass (year 2), Winter wheat (year 3), Maize (year 4)
2	Plant production	26.0	Spring Barley (year 1), Grass (year 2), Winter wheat (year 3 + year 4)
3	Pig farms	20.2	Spring Barley (year 1), Grass (year 2), Winter wheat (year 3), Winter barley (year 4)
4	Grass	25.1	Grass (year 1-4)
5	Coniferous forest	1.8	
6	Deciduous forest	8.1	
7	Urban area	7.5	50% impervious and 50% grass

Slope: The area was divided in 2 slope classes: (i) from 0% to 2% and (ii) higher than 2%.

The HRUs were created by overlaying soil, land use and slope maps. As a result, there were 808 HRUs in 30 sub-basins in the Odense River basin SWAT model.

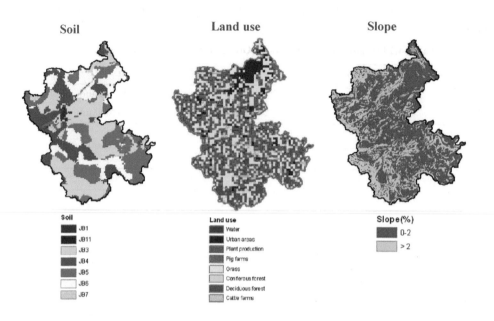

Figure 4.2 Soil, land use, slope maps which are overlaid to define HRUs

Locating climate stations

Figure 4.3 presents the input of climate data into the SWAT model. Daily precipitation and temperature data for the Odense River basin region in the form of 11 interpolated 10 km x 10 km precipitation grids and 4 interpolated 20 km x 20 km temperature grids, respectively, were obtained from the Danish Meteorological Institute. However, SWAT only can get the climate data from stations. Therefore, these available data were input to SWAT by creating 11 rainfall stations and 4 temperature stations located at the centroid of each precipitation or temperature grid (figure 4.3). Solar radiation, relative humidity and wind speed data were taken from a single weather station for the whole area (figure 4.3). Potential evapotranspiration was calculated by the Penman – Monteith method (Monteith, 1965; Penman, 1956) in this study.

Pollution sources

Due to the objective of modelling only nitrogen cycling and transport, only data related to nitrogen compounds were introduced into the model. Nitrogen loadings originate from point sources and fertilizer application. Nine point sources which represented discharges from sewerage systems were accounted for in the model. There were 2 types of fertilizer applied: (i) mineral fertilizer N50S which composes of 50.4% NH_4-N and NO_3-N as remainder and (ii) cattle slurry Cslurry13 and pig slurry Slurry13 which both include 40% of NH_4-N and 60% of organic N.

Precipitation stations

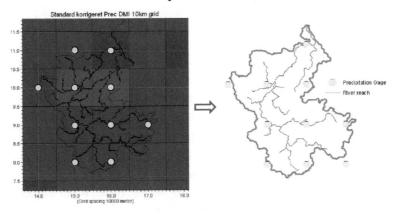

Temperature stations

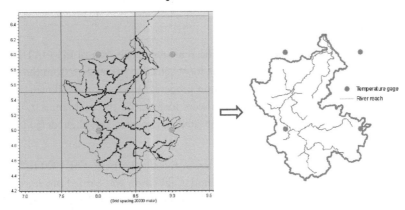

Solar radiation, humidity, wind speed

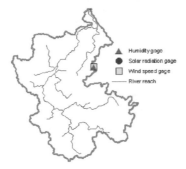

Figure 4.3 Locating climate stations in setting up SWAT models for the Odense river basin

Tile drainage

In SWAT, the parameters related to tile drainage simulation consist of d_{drain} (depth from soil surface to tile drains), dep_imp (depth to impervious layer), t_{drain} and g_{drain} (drain tile lag time). D_{drain} was set at 1 m in every HRU in agricultural areas which have the land use classification of cattle farms, plant production, pig farms or grass (table 4.2). SWAT calculates tile flow together with water percolation in the soil profile, thus, it is necessary that tile drains be located inside the soil profile. With d_{drain} equals to 1 m in a 1.5 m soil profile, the condition for tile-drain locations is satisfied for the Odense river basin.

Dep_imp was set at approximately 3.2 m for the whole basin to allow the rising of the perched water table which is necessary to generate the tile flow. If the groundwater table height exceeds the height of tile drains above the impervious zone, tile drainage will occur. Since there is no supporting data to identify the value of dep_imp parameter, it was assumed that the whole area has the same dep_imp value and this parameter was included in the calibration.

4.1.2 Calibration/ validation

A sensitivity analysis was implemented using the sensitivity analysis tool LH-OAT in SWAT, prior to performing the SWAT calibration. The most sensitive parameters from the sensitivity analysis results were considered in the calibration process. Moreover, the following additional parameters relating to tile drainage simulation were included in the calibration: t_{drain} and g_{drain} and dep_imp. Table 4.3 shows the list of calibrated parameters and their default and final values.

The SWAT model was run with a daily time-step in the 14-year period of 1990 to 2003. The first three years were used as a warming-up period. Calibration was carried out for the period of 1993 to 1998 while 1999 to 2003 served as the validation period. There are three gauging stations at which data for flow and nitrogen are available: 45_01, 45_21, 45_26 which correspond to outlets at sub-basin 17, 4 and 3, respectively (figure 4.4). The most downstream gauging station 45_26 corresponding to sub-basin 3 was chosen as calibrated station for flow and nitrogen fluxes. This station is located just downstream of the urban area, therefore, it is affected by not only outflow and pollutant fluxes from agricultural areas but also these from urban areas.

The SWAT flow results after calibration was also compared with measured data at gauging station 45_01 and 45_21 which correspond to outlets of sub-basin 17 and 4, respectively (figure 4.4).

Table 4.3 List of calibrated parameters and their default and final values

No.	Parameters	Description	Process	Default value	Calibrated value
Flow parameters					
1	esco	Soil evaporation compensation factor (-)	Evapo-transpiration	0.95	1.0
2	epco	Plant water uptake compensation factor (-)	Evapo-transpiration	1.0	0.05
3	cn2*	SCS runoff curve number for moisture condition II	Surface runoff	0	-0.28
4	surlag	Surface runoff lag time (days)	Surface runoff	4	22
5	alpha_bf	Baseflow alpha factor (days)	Groundwater flow	0.048	0.98
6	ch_n2	Manning's n value for main channel (-)	Channel routing	0.014	0.21
7	ch_k2	Effective hydraulic conductivity (mm/hr)	Channel routing	0	150
8	sol_awc*	Available water capacity of the soil layer (mm/mm)	Percolation	0	-0.24
9	sol_z*	Depth from soil surface to bottom of soil layers	Percolation	0	+0.16
10	dep_imp	Depth to impervious layer for modelling perched water tables (mm)	Tile drainage	6000	3200
11	tdrain	Time to drain soil to field capacity (hrs)	Tile drainage	0	25
12	gdrain	Drain tile lag time (hrs)	Tile drainage	0	24
Nitrogen parameters					
13	CMN	Rate factor for humus mineralization of active organic nitrogen	Mineralization	0.0003	0.0003
14	CDN	Denitrification exponential rate coefficient	Denitrification	1.4	0.07
15	SDNCO	Denitrification threshold water content	Denitrification	1.1	1
16	N_UPDIS	Nitrogen uptake distribution parameter	Nitrogen uptake	20	20
17	HLIFE_NGW	Half-life of nitrate in the shallow aquifer (days)	Groundwater nitrate	0	1

*: parameter that is changed by multiplying the original value and (1 + calibrated value)

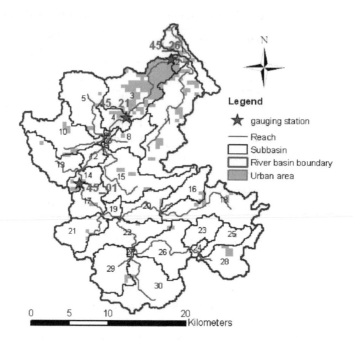

Figure 4.4 Gauging stations locating in the Odense river basin

To evaluate the SWAT performance for the simulated discharge and nitrogen flux, the following criteria were used:

- $NSE_{Q, daily\ or\ monthly}$: Nash-Sutcliffe efficiency calculated on the basis of observed and simulated daily or monthly discharge values (Nash and Sutcliffe, 1970).

- $NSE_{N, daily\ or\ monthly}$: Nash-Sutcliffe efficiency calculated on the basis of observed and simulated daily or monthly nitrate flux (Nash and Sutcliffe, 1970).

- $r_{Q, daily\ or\ monthly}$: the correlation coefficient between simulated and observed daily or monthly discharge values.

- $r_{N, daily\ or\ monthly}$: the correlation coefficient between simulated and observed daily or monthly nitrate flux.

The statistical results between the simulated constituents and corresponding measured values were also assessed according to criteria suggested by Moriasi et al. (2007) for judging successful model results.

4.1.3 Modelling results

4.1.3.1 Flow calibration

Table 4.4 and figure 4.5 show an overview of flow calibration results. The statistical evaluation of the SWAT daily streamflow predictions versus measured daily streamflow (table 4.4) resulted in NSE_Q values of 0.82 at station 45_26, the calibrated station, and 0.83 and 0.75 for stations 45_01 and 45_21, respectively. The corresponding NSE values for the validation period ranged from 0.75 to 0.80 across the three stations (table 4.4). The daily r_Q statistics all equalled to or were higher than 0.9 for both the calibration and validation periods. Figure 4.5a shows that SWAT accurately replicated much of the daily streamflow hydrograph at the calibrated station 45_26 (NSE_Q = 0.82 and 0.80 in the calibration and validation period, respectively) although some of the peak flows were under- or over-predicted and variations in low flows were not captured well. It can be seen from figure 4.5 that the measured daily streamflow hydrograph contains small fluctuations during the low flow periods. These small variations are possibly caused by tile drainage from low elevation areas where the groundwater table is very near the tile drain depths or from surface runoff in wetlands where the soil is already saturated with water or urban runoff.

The daily results were then aggregated into monthly time-steps to test the performance of the model for a longer time step. The NSE_Q statistics computed for the predicted monthly flow at ranged from 0.88 to 0.90 in the calibration period and from 0.79 to 0.90 in the validation period (table 4.4 and figure 4.6). The performance of SWAT was better for the monthly streamflow predictions as compared to the daily predictions, which is consistent with previous overall reviews of dozen of SWAT studies (Douglas-Mankin et al., 2010; Gassman et al., 2007; Tuppad et al., 2011). However, all of the daily and monthly statistics exceeded the criteria of "very good simulation results" proposed by Moriasi et al. (2007), assuming their criteria of NSE equal 0.75 for monthly streamflows is adaptable to daily comparisons as well.

Table 4.4 Performance criteria for the SWAT model in daily and monthly time-steps

Period	Station / Criteria	Daily			Monthly		
		45_26	45_21	45_01	45_26	45_21	45_01
Calibration	NSE_Q	0.82	0.75	0.83	0.90	0.88	0.90
	r_Q	0.92	0.90	0.92	0.97	0.96	0.95
Validation	NSE_Q	0.80	0.75	0.77	0.90	0.84	0.79
	r_Q	0.92	0.90	0.92	0.96	0.95	0.95

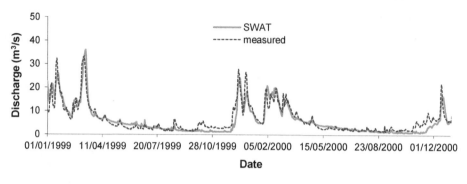

(a) Station 45_26 (outlet of sub-basin 3)

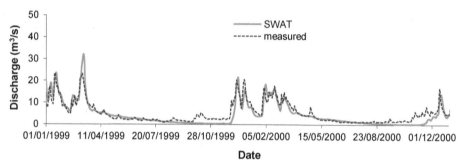

(b) Station 45_21 (outlet of sub-basin 4)

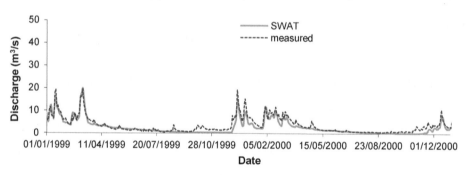

(c) Station 45_01 (outlet of sub-basin 17)

Figure 4.5 Comparison of discharge values from SWAT and measured data at three stations (45-26, 45_21 and 45_01) in the validation period

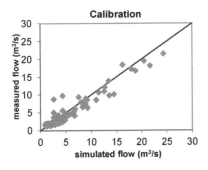
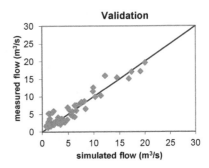

Figure 4.6 Scatter plots of monthly simulated and observed discharge in calibration and validation periods at the station 45-26

Breaking down average flow components for the period 1993-1998, SWAT predicted that tile drainage and groundwater flow share to be two dominant flow components (table 4.5). Surface runoff contributes only 52 mm while lateral flow is the smallest component with only about 9 mm. This result is compatible with the study of Dahl et al. (2007) who evaluated the flow path distributions and identified the dominant flow path through the riparian area to the stream for the Odense river basin based on the field work of Banke (2005) at seven transects along two main streams of the Odense river basin. Dahl et al. (2007) found that the dominant flow path is tile flow at five transects and groundwater flow at the other two. Surface runoff was found at four transects but was not the dominant flow path.

The temporal distribution of flow components for the whole basin is illustrated in figure 4.7. It can easily be observed that in the Odense river basin, it rains all the year, even in the summer period. However, because of higher temperature which causes much higher evaporation and relatively lower soil moisture in summer periods, the flow is lower in summer periods (April to October) than winter periods (November to March). Surface runoff is generated throughout the whole year when there is high precipitation; however, it is much higher in winter than summer periods corresponding to the huge difference in soil moisture between the two periods. Different from surface runoff, tile flow only appears in winter/high flow periods and is absent in summer/low flow periods. The reason is that evapotranspiration is high in summer; therefore, SWAT predicted less water percolation out of the soil profile and in turn the water table rarely reaches the tile drain level. Unlike the other two components, groundwater which is considered as a slow component contributes to total flow all the time. In winter, groundwater flow contributes to total flow together with surface runoff and tile flow while it is the dominant source in summer.

Table 4.5 Annual water balance in the SWAT model for Odense river basin from 1993-1998

Water balance components	mm water
Precipitation	852
Surface runoff	52
Lateral flow	9
Tile drainage	156
Groundwater flow	160
Evapotranspiration	461

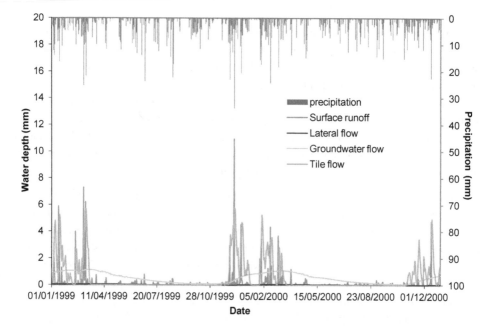

Figure 4.7 Temporal distribution of different flow components for the whole basin in the 1990-2000

4.1.3.2 Nitrogen calibration

In SWAT, ammonium is assumed to be easily absorbed by soil particles, thus, it is not considered in the nutrient transport. Nitrate, which is very susceptible to leaching, can be lost through surface runoff, lateral flow, tile flow and groundwater flow. Therefore, in this section, the calibration only focuses on nitrate fluxes.

✓ *Sensitivity of nitrogen related parameters*

Figure 4.8 shows the sensitivity of nitrogen-related parameters on the average daily NO_3 load at the gauging station 45_26. There are five parameters that can affect the NO_3 flux simulation: (1) *NPERCO* (nitrate percolation coefficient) controls the amount of nitrate following surface runoff, (2) *CDN* (denitrification exponential rate coefficient) controls the amount of nitrate removed from the basin through denitrification, (3) *CMN* (rate factor for humus mineralization of active organic nitrogen) affects the mineralization from organic N to nitrate, (4) *N_UPDIS* (nitrogen uptake distribution parameter) decides whether top layers or all layers of soil are favoured for nitrate uptake and (5) *HLIFE_NGW* (half-life of nitrate in the shallow aquifer) represents for all reactions that occur to nitrate stored in the shallow aquifer. It is clearly seen in figure 4.8 that *CDN* is the most sensitive parameter that significantly affects the average daily NO_3 load followed by the parameter *CMN*. *CMN* seems to have considerable impact when its value is less than 0.01. The *HLIFE_NGW* has higher effect when its value is less than 10. The remaining three parameters have no or very little effect on the NO_3 flux results. Therefore, *CMN* and *CDN* and *HLIFE_NGW* were considered in the nitrate calibration. Their calibrated values are shown in table 4.3.

Besides these parameters, there is a very important parameter that indicates whether or not denitrification process occurs, *SDNCO* (denitrification threshold water content). *SDNCO* is defined as the fraction of field capacity above which denitrification will occur. For the application of SWAT on Odense river basin, if *SDNCO* >1, denitrification does not happen. Therefore, *SDNCO* was set at 1 to allow denitrification occurring, which means that denitrification will occur if soil water is higher than field capacity.

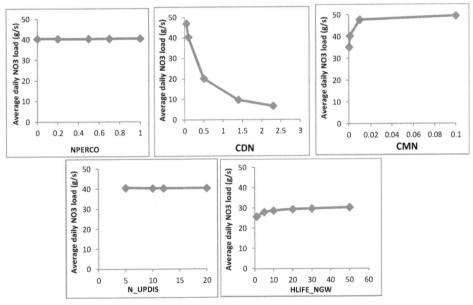

Figure 4.8 Sensitivity of nitrogen-related parameters on average daily NO_3 load at the station 45_26

Nitrate is discharged primarily from point sources during low flow periods in the Odense River basin, but originates mostly from leaching in agricultural areas during high flow periods and hence depends heavily on the river basin flow and nitrogen cycling and transport processes. The comparison of daily nitrate flux simulated by SWAT and measured data can be found in figure 4.9. It can be observed that the magnitude and the trend of the nitrogen flux predicted with SWAT were quite similar to the measured data. However, SWAT did not accurately capture many of the daily fluxes well, especially several of the peak nitrate fluxes measured during the high flow periods. One possible reason for this weak result was that the permutation of crops was not implemented in SWAT, which affected the simulated inputs including the amount of fertilizer applied and the time of application. In addition, inaccurate simulations of the different flow components as well as the nitrogen cycling processes also likely contributed to problems in replicating the daily nitrate fluxes. However, regarding to the performance criteria, the daily NSE_N value computed between SWAT and the measured data was 0.55, which was considered as "satisfactory" following the criteria suggested by Moriasi et al. (2007) of 0.5 (although that was suggested for monthly time step comparisons).

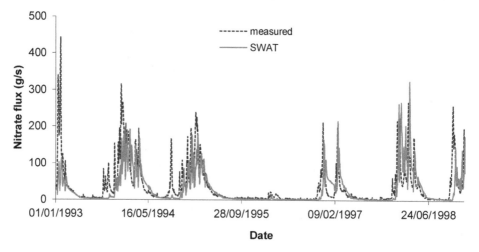

Figure 4.9 Comparison of nitrogen flux between SWAT model and measured data at the station 45_26

Figure 4.10 shows the performance of SWAT in modelling the nitrate flux at both daily and monthly time-steps, relative to 1:1 line plots and the resulting NSE_N and r_N statistics. It can clearly be seen that the monthly NSE_N value of 0.69 was stronger than the daily NSE_N value and would be considered "good" for monthly time step results according to the model evaluation guidelines of Moriasi et al. (2007). These results coupled with the similarity in magnitude and variation of the predicted versus measured daily nitrogen fluxes imply that SWAT can be an effective tool for simulating nitrogen loadings in the Odense River basin.

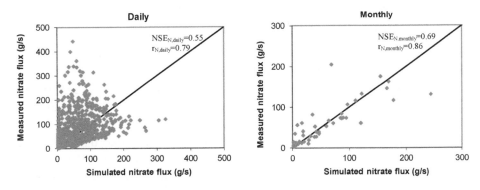

Figure 4.10 Scatter plots of daily and monthly simulated and observed nitrate flux at the station 45-26

The annual nitrate balance for the period 1993-1998 is shown in figure 4.11. The primary sources for nitrate are from inorganic fertilizer (49.6 kg/ha), mineralization from humus and fresh residue (101.5 kg/ha) and nitrification from ammonia (91.2 kg/ha). Nitrate was permanently removed by denitrification (13.4 kg/ha) or by some biological/chemical processes in the aquifer (7.3 kg/ha). Nitrate was also uptaken by plant (189.1 kg/ha) a part of which was removed from the basin through crop harvesting while the rest stayed in the basin and was decomposed later. Nitrate followed different flow components to the streams in which tile flow and groundwater flow were the most significant sources of nitrate to the streams with 20.5 and 8.9 kg/ha, respectively. Although groundwater and tile flow share to be dominant sources of flow to the river, tile flow is the dominant path for nitrate fluxes.

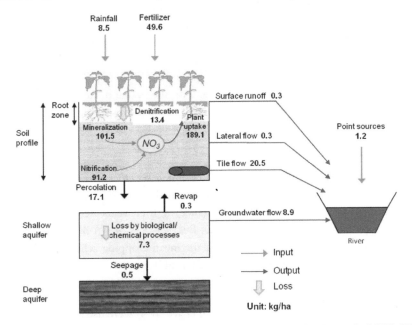

Figure 4.11 Annual nitrate balance for the Odense river basin for the period 1993-1998

4.2 DAISY-MIKE SHE MODEL FOR THE ODENSE RIVER BASIN

4.2.1 DAISY-MIKE SHE model setup for Odense river basin

The Odense river basin was already built using the DAISY-MIKE SHE model in several versions. The previous DAISY-MIKE SHE-based studies reported by Van der Keur et al. (2008) and Hansen et al. (2009) were both based on an earlier application of DAISY-MIKE SHE described by Nielsen et al. (2004). The Odense River basin served as the study area for the DAISY-MIKE SHE application described by Van der Keur et al. (2008). However, the larger Odense Fjord basin, which covers an area of 1,312 km^2 (including the fjord area) and encompasses the Odense River basin, was the area simulated with DAISY-MIKE SHE in the study by Hansen et al. (2009). Nevertheless, the research performed by Hansen et al. (2009) provides a suitable basis of comparison in this thesis due to the similar characteristics between the two basins. With the purpose of comparing the performance of SWAT and DAISY-MIKE SHE models in the Odense River basin case study (the comparison is described in chapter 5), the input data of the two models were kept similar or comparable. Therefore, the description of the DAISY-MIKE SHE model components below focused only on important differences in the set-up of DAISY-MIKE SHE model relative to SWAT.

4.2.1.1 DAISY model

✓ Boundary conditions

There are three different lower boundary condition options in DAISY: a constant groundwater level, a gravitational gradient and a time-varying groundwater component using a drain pipe option in DAISY. In the model of the Odense River basin, the lower boundary conditions were assigned based on the simulated water table from the Danish National Water Resource Model (Henriksen et al., 2003). The drain pipe option was applied at a depth of 1 meter in all DAISY columns where the simulated mean water table depth is shallow. The lower boundary condition was set to the deep groundwater level option in all columns where the average water table is deeper than 3 m. For the wetland area, the lower boundary condition was a fixed water table approximately 45 cm below the ground (Hansen et al., 2009).

✓ Crop rotation

Different from the SWAT model set-up, the cropping schemes in DAISY related to cattle farms, plant production and pig farms were permutated to ensure that each crop was equally represented for each climate year. For instance, a four years scheme with A–B–C–D successive crops has permutations A–B–C–D, B–C–D–A, C–D–A–B, and D–A–B–C. The permuted outputs were averaged to represent the mean cropping scheme in the agricultural land use which was then used as input to the MIKE SHE model.

✓ Calibration

Standard parameter values mostly from Styczen et al. (2004) were used. It was not possible to calibrate the simulated percolation of individual DAISY columns. However, Hansen et al.

(2009) made some manual calibrations of the parameters after evaluation of catchment simulations in an attempt to improve the performance of the simulated discharge at the catchment scale from Nielsen et al. (2004). Simulated nitrogen balances were calibrated against agricultural crop yield statistics in the period 1991–2000.

4.2.1.2 MIKE SHE model

✓ Climate

Different from the SWAT model, MIKE SHE can accept climate data represented as grids. In this model, 14 interpolated 10 km x 10 km precipitation grids and 4 interpolated 20 km x 20 km temperature grids from the Danish Meteorological Institute were used. A global radiation value from a single observation point in the central part of the catchment was applied. The reference evapotranspiration, which is used for calculation of the actual evapotranspiration in the DAISY model, was calculated by a modified version of Makkink's Equation (Hansen, 1984).

✓ Hydro-geological model

The hydro-geological model for the Odense river basin is characterized by 9 geological layers. The top layer is characterized as fractured till, while the succeeding layers (2–8) are of alternating aquitard (till) and aquifer (sand) material, starting and ending with an aquitard. Sandy units in aquitards are included as sand lenses in the geological model. The lower ninth layer constitutes Palaeocene marl and clays and older limestone. The model is discretised into 500m x 500m grid blocks (Hansen et al., 2009). MIKE SHE simulates saturated flow using the 3D Boussinesq equation (Boussinesq, 1872).

✓ Tile drainage

Tile drainage was modelled using the built-in drain routing option in MIKE SHE. If the groundwater level exceeds a specified threshold height called drain level (1 m below the surface in this study), the excess water is routed to the nearest river reach by a first order rate specified by the drain time constant (s^{-1}).

✓ Delineation of redox interface

In the saturated zone, it is assumed that nitrate is reduced very slowly in layers above the location of the redox interface whereas all nitrate transported to layers below the interface is removed instantaneously. In order to account for this, oxidised and reduced zones were introduced in MIKE-SHE, which are separated by a redox interface and have very high and very low nitrate half-life parameters, respectively. A schematization of the relationship between the redox interface and nitrate movement is depicted in figure 4.12. A half-life of 2 years was applied for oxidized zone after calibration. It was assumed that the depth to the redox interface is related to soil types at 1 meter below the surface. Sandy areas are assumed to have higher infiltration rates than more clayey areas and therefore deeper redox interfaces. The redox interface was assumed to be located 2 m below the surface in clay and organic soil areas, and 3.5 m or 8 m below the surface in till areas below and above an elevation of 45 m, respectively. These divisions represent a distinction between areas with shallow and deep groundwater tables. A deeper unsaturated zone will result in a deeper redox interface, owing to faster diffusive transport of oxidizing agents, especially oxygen, above the water table (Hansen et al., 2009). Finally, the redox interface in sandy areas was set to 16 m below the surface.

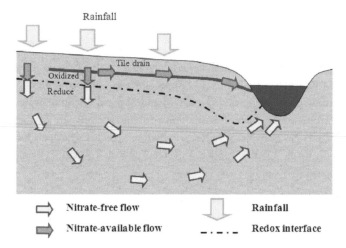

Figure 4.12 Schematisation of the relationship between redox interface and nitrate movement

4.2.1.3 Coupling of MIKE-SHE and DAISY

A total of 6061 DAISY column simulations were made to model the water and nitrogen budget of the root zone. Results from DAISY columns were then distributed to corresponding field blocks. Subsequently, outputs from field blocks were then aggregated as daily net values and distributed to MIKE SHE grid blocks using a weighting area procedure. MIKE SHE used the outputs of water percolation and nitrogen leaching in simulating processes in saturated zone.

This model was run using the same parameters as used by Henriksen et al. (2003). Only the drain time constant was recalibrated. The performance of DAISY-MIKE SHE on flow and nitrogen simulations is shown in table 4.6.

4.2.2 Modelling results

✓ Root zone model

After calibration of the crop modules, the simulated harvested nitrogen from the represented crops was within the range of -39% to +11% of the statistical measures of harvested nitrogen from the same crops (Nielsen et al., 2004). Average measured harvested yields for the crops are available from statistics for Funen Island and the nitrogen content in the harvested crops is based on national average values.

✓ Catchment model

The difference between annual observed and simulated discharge (ME_Q) were negative for all three stations in both calibration and validation periods. That means the simulated results underestimated the actual discharge. The underestimation was more significant in validation periods (ME_Q is from -11 to -6 %) compared to calibration periods (ME_Q is from -5 to -2 %). The NSE_Q ranged from 0.51 to 0.63 in calibration periods and 0.45 to 0.58 in validation periods. The model results were evaluated as satisfactory by the model developers. According to the performance criteria from Moriasi et al. (2007), these results are also considered as satisfactory assuming their criteria for monthly streamflow is also adaptable to daily streamflow. However, the low Nash-Sutcliffe efficiency (NSE_Q) is explained by the underestimation of daily discharge during autumn and the overestimation during wet winter and spring periods as well as a bias in overestimation in wet years and underestimation in dry years (Hansen et al., 2009).

Table 4.6 Performance criteria for flow and nitrogen simulation from the DAISY-MIKE SHE model (Hansen et al., 2009)

Flow calibration							
Station	Calibration period (1991-2000)				Validation period (2001-2002)			
	Observed	ME_Q	ME_Q	NSE_Q	Observed	ME_Q	ME_Q	NSE_Q
	mm/yr	mm/yr	%			mm/yr	%	
45_01	320	15	-5	0.60	354	38	-11	0.58
45_21	297	7	-2	0.51	336	34	-10	0.45
45_26	329	8	-2	0.63	354	21	-6	0.54
Nitrate simulation								
Station	Calibration period (1998-2000)				Validation period (2001-2002)			
	Observed	ME_N	ME_N	NSE_N	Observed	ME_Q	ME_Q	NSE_N
	tonnes/yr	tonnes/yr	%		tonnes/yr	tonnes/yr	%	
45_21	1291	38	-3	0.21	1009	222	-22	0.57

ME_Q: the difference between observed and simulated discharges averaged over the calibration and validation periods.

ME_N: the difference between observed and simulated fluxes of total nitrogen averaged over the calibration and validation periods.

According to nitrate calibration, the model only underestimated 3% of the annual nitrate load in the calibration period, but predicted 22% less than the actual load in the validation period. However, the validation period is very short and limited in data; therefore, this result may not be accurately represented the model performance. The low Nash- Sutcliffe efficiency for nitrate (0.21) indicated that the predictive capability of the model for simulating seasonal and inter-annual dynamics is somewhat limited (Hansen et al., 2009)

Chapter 5

COMPARISON AND EVALUATION OF MODEL STRUCTURES FOR THE SIMULATION OF FLOW AND NITROGEN FLUXES IN A TILE-DRAINED RIVER BASIN

5.1 INTRODUCTION

In this chapter, different models with different model structures are compared and evaluated for the simulation of flow and nitrogen fluxes. The applied case study is the Odense river basin which is a typical tile-drained river basin serving for agricultural purposes. Two main comparisons were implemented: (i) comparison between SWAT and DAISY-MIKE SHE and (ii) comparison between different SWAT models with different setups. The corresponding objectives of this chapter are (i) evaluating the performance differences of two distributed river basin models which use different concepts in modelling flow and nitrogen and (ii) assessing the importance of model structures in setting up a model for a real case study.

First, the SWAT model for the Odense river basin which was described in chapter 4 was compared with the existing DAISY-MIKE SHE model in terms of flow and nitrate fluxes. Previously, El-Nasr et al. (2005) performed a hydrological comparison between SWAT and MIKE SHE for the Jeker River basin in east central Belgium. The comparison showed that both models were able to simulate the hydrology in an acceptable way although the overall variation of the river discharge was predicted slightly better by MIKE SHE which may have been due to the fact that the aquifer system could not be modeled adequately in the SWAT model. However, the authors did not analyze the difference between flow components. Nasr et al. (2007) also compared SWAT with the Système Hydrologique Européen TRANsport (SHETRAN) model (including the use of grid oriented phosphorus component or GOPC), which was also derived from the original SHE model and is closely related to MIKE SHE (Refsgaard et al., 2010), and a third water quality model for three river basins in Ireland. They reported that the SWAT simulations resulted in the strongest daily calibration of total phosphorus loads but that annual exports of total phosphorus were simulated at similar levels of accuracy by all three models. However, no previous study has compared SWAT and DAISY-MIKE SHE simulations of flow and nutrient loadings, including accounting of flow and nitrate transport via subsurface tile drains. Thus in this thesis, both SWAT and DAISY-MIKE SHE were evaluated to assess the suitability of the different approaches used in the two models for the Odense River basin in Denmark. The Odense River basin is dominated by extensive agricultural production areas and over 50% of the agricultural areas are drained by subsurface tile drains, providing an ideal system for comparing both the hydrological and nitrate transport components incorporated in SWAT and DAISY-MIKE SHE.

Using the SWAT model that was described in chapter 4, SWAT simulations were conducted, which provided overall estimates of streamflow and nitrate fluxes at the basin outlet including flow and nitrate contribution routed via subsurface tile drains. These SWAT outputs were then compared with existing DAISY-MIKE SHE simulations of the Odense River basin performed by van der Keur et al. (2008) and the larger Odense Fjord basin reported by Hansen et al. (2009), which also included estimates of tile drain contributions of flow and nitrate loads to the respective river basin outlets, to assess the overall performance of the two models.

Secondly, besides comparing two models with different approaches, different setups using the SWAT model were also built and compared to evaluate the flow and nitrogen performance with different structures. There are two different setups of the Odense river basin built on SWAT with different structures: (i) without tile drains, and (ii) tile drains applied. The comparison was implemented to verify the two hypotheses below:

✓ Odense River basin is a lowland river basin with intensively drained agriculture areas. Therefore, tile drainage is assumed to have a strong impact on lowland hydrology. The inclusion of tile drainage in the model is supposed to obtain a better representation of hydrological processes in the case study.

✓ Previous studies show that many different parameter sets can give almost identical fits to the measured data (the equifinality problem). The two models were calibrated using auto-calibration to see whether two different model structures can both achieve a good fit to the measurements.

5.2 COMPARISON BETWEEN SWAT AND DAISY-MIKE SHE IN FLOW AND NITROGEN SIMULATIONS

The results of the SWAT model were compared with the DAISY-MIKE SHE model in terms of discharge and nitrogen simulation, following the completion of the SWAT calibration and validation, based on the DAISY-MIKE SHE results reported by Van der Keur et al. (2008) and Hansen et al. (2009). Van der Keur et al. (2008) simplified the original DAISY-MIKE SHE model set-up reported by Nielsen et al. (2004) to perform an uncertainty analysis of discharge and nitrate loadings, by varying identified model parameters that most likely contributed to uncertainty in discharge and nitrate loadings. The model parameters included in the uncertainty analysis were: (1) soil hydraulic properties, which are related to transport of nitrate from the root zone, (2) the slurry application amount, which affect the nitrogen turnover processes, and (3) root depth, which contributes to uncertainty related to the soil water balance. Van der Keur et al. (2008) then performed a total of 24 DAISY-MIKE SHE runs, which were implemented with changing parameter sets generated using a Latin Hypercube Sampling technique. The results from the SWAT simulation were compared with the maximum and minimum results of the 24 DAISY-MIKE SHE runs at gauging station 45_26 at a daily time-step during the period 1993 to 1998. This provided a basis of comparison between SWAT and DAISY-MIKE SHE for both flow and nitrogen simulations, while taking into account the uncertainty in soil hydraulic and slurry parameters.

The additional comparison between the SWAT results and DAISY-MIKE SHE results reported by Hansen et al. (2009) focused primarily on general water balance differences predicted between the two studies. Direct comparisons between the two studies are limited due to differences in the size of the basin regions, land use distributions, precipitation inputs and other study characteristics. However, the comparison does provide important insights into differences in flow partitioning and other responses between the two models, as discussed in more detail in below sections.

5.2.1 Comparison in flow simulation

Figure 5.1 shows the SWAT-predicted streamflow at gauging station 45_26 versus the maximum and minimum streamflow time series values of the 24 DAISY-MIKE SHE simulations reported by van der Keur et al. (2008). It can be observed that the SWAT results fit quite well within the range of the DAISY-MIKE SHE values. In the high flow period, almost all the SWAT values were within the range but 36% of the SWAT flow values were smaller than the corresponding DAISY-MIKE SHE minimum values. Most of the lower SWAT flow predictions occurred during the low flow periods; however, the difference in values was small. The measured data also had 39% of the values below the lower range which happened in the low flow period (table 5.1).

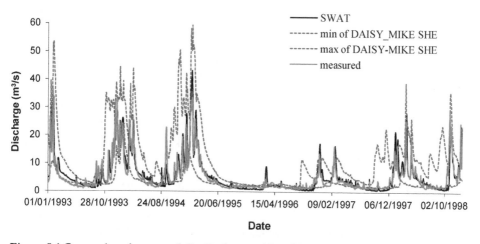

Figure 5.1 Comparison between daily discharge of SWAT, measured data and the range of discharge value from the DAISY-MIKE SHE model at the station 45-26

Table 5.1 Comparison between simulated results from SWAT and measured data with the range of simulated results from DAISY-MIKE SHE

	Percentage of flow values			Percentage of nitrate flux values		
	< min[a]	Between min and max	> max[b]	< min	Between min and max	> max
SWAT	38	57	5	46	38	17
Measured data	39	55	6	38	43	19

[a] *min: minimum value in each time-step of 24 simulations from DAISY-MIKE SHE model*

[b] *max: maximum value in each time-step of 24 simulations from DAISY-MIKE SHE model*

The hydrograph of the SWAT model was also compared with the 50th percentile (median) ranked DAISY-MIKE SHE outputs from the 24 simulations (figure 5.2). When comparing the two models to each other, the correlation coefficient was 0.86 for the period 1990-2000, which implied good correspondence between the two models. However, the SWAT model replicated the measured streamflow hydrograph more accurately than DAISY-MIKE SHE, which was also confirmed by the higher NSE and correlation coefficient statics computed for the SWAT results (table 5.2).

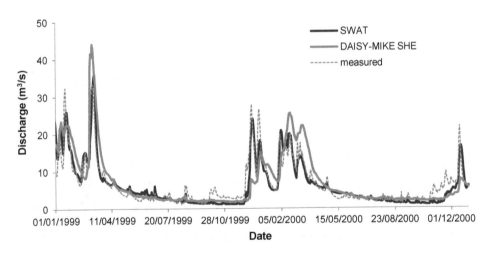

Figure 5.2 Comparison of simulated daily discharge between SWAT model, median discharge of DAISY-MIKE SHE and measured data

Table 5.2 Performance criteria for SWAT and DAISY-MIKE SHE

Criteria	SWAT	DAISY-MIKE SHE
$NSE_{Q,daily}$	0.80	0.62
$r_{Q,\ daily}$	0.92	0.82

An annual average water balance comparison (table 5.3) was also performed between SWAT and the DAISY-MIKE SHE results reported by Hansen et al. (2009), who simulated the larger Odense Fjord basin as previously described. It is noted that the figures in table 5.3 is the overall amount of flow that the streams receive from all sub-basins, not the streamflow at the outlet of the river basin. With the lag time for flow routing, the overall flow at the outlet of the river basin is certainly smaller than these values. From table 5.4, striking water balance differences were reported in the two studies, with small amount of surface runoff (51 mm), and nearly equal amounts of tile flow (167 mm) and groundwater flow (173 mm) predicted by SWAT versus essentially no surface runoff (2 mm), over 200 mm of tile flow, and very little groundwater flow for the DAISY-MIKE SHE simulation. There was a big difference in the overall streamflow between the two models. It is possibly due to some differences in inputs and study area. The DAISY-MIKE SHE model from Hansen et al. (2009) was built for the 1312 km^2 Odense Fjord catchment and based on point raingauge data, while the SWAT model was built for the 622 km^2 Odense river catchment based on 10 km grid precipitation data. However, we still can compare the water balances between the two models by looking at their flow component breakdowns. Virtually all of the subsurface flow occurred as tile flow in DAISY-MIKE SHE while subsurface flow inputs were roughly evenly split between tile flow (167 mm) and groundwater flow (173 mm) in SWAT. Overall, SWAT predicted about 200 mm of additional streamflow to the Odense River as compared to the streamflow predicted with DAISY-MIKE SHE by Hansen et al. (2009), due to the higher surface runoff predicted by SWAT and relatively high amount of leaching to groundwater, etc. in DAISY-MIKE SHE (which did not occur in SWAT). Based on the field work by Banke (2005) at seven transects along two main streams of the Odense river basin, Dahl et al. (2007) evaluated the flow path distributions and identified the dominant flow path through the riparian area to the stream. They found that the dominant flow path is tile flow at five transects and groundwater flow at the other two. Surface runoff was found at four transects but was not the dominant flow path. This would indicate that both SWAT and DAISY-MIKE SHE accurately identify subsurface flow to be the main contributor to streamflow, however, the proportion between tile flow and groundwater flow are different between the two models. As mentioned in chapter 4, SWAT predicted tile flow to only occur in high flow periods and groundwater is the main source to streamflow in low flow periods. It is due to high evapotranspiration in low flow periods resulting in less water percolation out of the soil profile and low water table that rarely reaches the tile drain level. Different from SWAT, DAISY-MIKE SHE still predicted tile flow to be generated as the main source for low flow periods. This difference in tile drainage occurrence between the two models is illustrated in figure 5.3 for the period April – November of 1999-2000 which is within the validation period. This issue will be discussed later in the Discussion section.

Table 5.3 Annual water balance in the SWAT and DAISY-MIKE SHE models in the period of 1998-2002

Water balance components	SWAT (mm)	DAISY-MIKE SHE (Hansen et al., 2009)* (mm)
Precipitation	935	Input to DAISY
Surface runoff	51	2
Lateral flow	10	-
Tile drainage	167	203
Groundwater flow	173	9
Evapotranspiration	517	Output from DAISY
Sink (groundwater abstraction, flow across boundaries, storage change)	17	66

* This water balance was made for the 1312 km² Odense Fjord catchment and based on point raingauge data, while the SWAT water balances were made for the 622 km² Odense river catchment based on 10 km grid precipitation data.

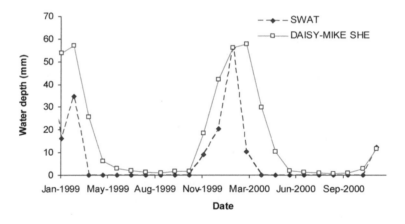

Figure 5.3 Comparison of tile drainage values between the SWAT and DAISY-MIKE SHE models in the validation period

5.2.2 Comparison in nitrogen simulation

The SWAT-predicted nitrate flux at gauging station 45_26 was also compared with the maximum and minimum time series of the 24 DAISY-MIKE SHE model simulations (van der Keur et al., 2008), similar to the previously described flow comparisons. It can be seen in figure 5.4 that the SWAT results fit quite well within the range between maximum and minimum values from the DAISY-MIKE SHE model during high flow periods although there are some peaks in 1997 and 1998 exceeding the maximum values from DAISY-MIKE SHE. Because of the high amount of water leaching in the soil profile and shallow aquifer in high flow periods, the nitrogen processes in the soil and shallow aquifer become very important for nitrate flux simulation. Therefore, these results imply that SWAT simulates river basin nitrogen processes reasonably during high flow periods, especially considering the fact that the parameter uncertainty accounted for in the 24 DAISY-MIKE SHE simulations. Similar to the flow comparisons, 46% of the SWAT-estimated nitrate fluxes were smaller than the corresponding minimum DAISY-MIKE SHE values and most of those lower predicted values occurred in the low flow periods (table 5.1) due to the lower flows predicted by SWAT.

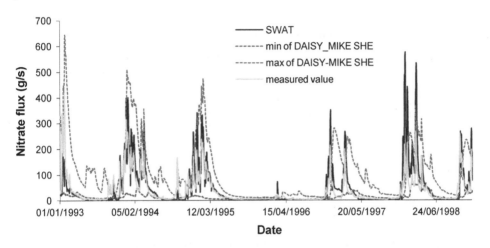

Figure 5.4 Comparison between nitrate flux of SWAT with tile drainage, measured value and the range of nitrate loads simulated from DAISY-MIKE SHE model at the station 45_26

Table 5.4 shows the result of comparing the annual average nitrate fluxes at the outlet of the river basin predicted with SWAT, versus similar values estimated with DAISY-MIKE SHE by Hansen et al. (2009) for the larger Odense Fjord basin, for four pathways: surface runoff, tile flow, groundwater flow and lateral flow. Both model can be rated as "satisfactory", based on the SWAT statistical results reported here and previous DAISY-MIKE SHE testing described by Hansen et al. (2009), but the nitrate flux pathways varied considerably between the two model studies. The predicted surface runoff nitrate fluxes were very small in both studies due to the very small surface runoff generated by DAISY-MIKE SHE (table 5.3) and incorporation of inorganic fertilizer beneath the soil surface in SWAT. In contrast, the predicted tile flow and groundwater flow nitrate fluxes were very different between the two models (table 5.4). In DAISY-MIKE SHE, the majority of nitrate flux occur via tile-drain flow due to the dominant tile-drain flow contributions (table 5.3), and possibly because most of the nitrate that goes through the groundwater passes the redox line and then becomes denitrified. Similar to DAISY-MIKE SHE, tile-drain nitrate flux was also the dominant contribution to the river although tile flow and groundwater flow shared to be the dominant flow paths. Nitrate fluxes via tile drainage were comparable between the two models while nitrate fluxes via groundwater flow had a big difference. SWAT predicted much more nitrate fluxes following groundwater flow than DAISY-MIKE SHE because of the higher amount of groundwater flow predicted by SWAT. Another possible reason is the difference in nitrate removal concepts simulating in the two models. DAISY-MIKE SHE assumes that all nitrate fluxes passing the redox line is denitrified immediately, thus, this process occurs very fast and can significantly decrease the nitrate fluxes via groundwater flow. On the other hand, SWAT simulates the nitrate removal processes in shallow aquifer based on a half life parameter; therefore, this process cannot happen immediately but takes some time for nitrate fluxes to be decreased.

Table 5.4 Comparison of nitrate fluxes between the SWAT and DAISY-MIKE SHE model

Nitrate fluxes out of the basin	SWAT (kg/ha)	DAISY-MIKE SHE (Hansen et al., 2009) (kg/ha)
- Through surface runoff	0.3	0.04
- Through lateral flow	0.3	0
- Through tile flow	20.5	21.8
- Through groundwater flow	8.9	0.6
Loads to the river from diffuse sources	**30.0**	**22.5**

5.3 COMPARISON BETWEEN DIFFERENT SWAT SETUPS IN FLOW AND NITROGEN SIMULATION

5.3.1 Descriptions of different SWAT setups

Two different model setups were carried out to test the role of tile drainage in the simulation of hydrological balance

- Setup (1): tile drainage not included

- Setup (2): with tile drains applied - same as the model described in chapter 4

5.3.2 Comparison between different setups

The comparisons between different setups were conducted to reach the two objectives. Figure 5.5 illustrates the methodology of these comparisons and their objectives.

✓ To test the influence of tile drainage inclusion to the SWAT model, setup (1) without tile drainage was used as a reference model to compare with setup (2) in which this component was added. Sensitivity analysis and manual calibration were carried out for setup (1) to get an optimized parameter set. This parameter set was applied for setup (2) to ensure that two models are comparable. Setup (2) was then compared with setup (1) to study the difference that the tile drainage component caused in flow and nitrate simulations. This methodology was proposed and used by Kiesel et al. (2010).

✓ To test the model performance in different structures with the help of auto-calibration, the two models were calibrated with the same parameter list (table 4.3 in chapter 4). It is aimed at verifying if auto-calibration can help the model to compensate the lacking processes by increasing or decreasing the effect of other processes.

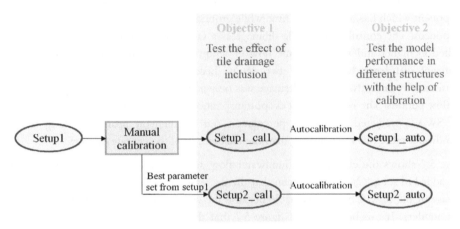

Setup()_cal1: SWAT setup that uses the best parameter set from setup (1)

Setup()_auto: SWAT setup that uses the best parameter set from autocalibration procedure

Figure 5.5 Methodology of comparing different SWAT setups

5.3.3 Results

Setup (1) without tile drainage inclusion was calibrated manually to get the reasonable trend and magnitude compared to measurements (figure 5.6a). NSE_Q and r_Q for daily results for setup (1) after manual calibration is 0.62 and 0.82, respectively. The results are considered satisfactory according to Moriasi et al. (2007). The parameter set from manual calibration for setup (1) was also applied in setup (2) to ensure a fair comparison between the two models and an accurate assessment on the effect of tile drainage on the SWAT model for Odense river basin.

5.3.3.1 The effect of tile drainage inclusion

From figure 5.6a and 5.6b, it is clearly observed that setup (2) fitted better to measurement than setup (1). Performance criteria, NSE_Q and r_Q increased respectively from 0.59 and 0.79 in setup (1) to 0.72 and 0.87 in setup (2) (table 5.5). The difference between the two models was clearer in high flow periods (November to March) than low flow periods (April to October). In high flow periods, the inclusion of tile drain parameters improved the SWAT model performance by producing higher peak flows and steeper hydrograph recession (figure 5.6). In setup (1), groundwater contributed the most significant amount of flow while groundwater and tile flow shared flow contribution in setup (2). Tile flow is a fast flow component which has a shorter lag time while groundwater flow is considered as a slow flow component. The contribution of tile drainage as a fast flow component in the flow periods resulted in higher and more dynamic flow in setup (2). In low flow periods, there was no significant difference between the two setups because flow in both setups was mostly contributed by groundwater. Tile drainage was only generated in the high flow periods, not in low flow periods. The reason is that evapotranspiration is high in low flow periods (summer); thus SWAT predicted less water percolation out of the soil profile resulting that the groundwater table was rarely higher than the tile drain level to generate tile flow.

Figure 5.7 shows the change of groundwater flow and tile flow when tile drain parameters were added in setup (2). It can be seen that the decrease of groundwater came along with the increase of tile flow. This is reasonable because the source of tile flow is from groundwater from aquifers. It can be noticed in figure 5.7 that the decrease of groundwater was higher than increase of tile flow. The reason is that not all groundwater loss became tile flow; some contributed to the change of soil moisture which resulted in small change in surface runoff and lateral flow.

Table 5.5 Performance criteria for different SWAT setups at the gauging station 45_26 using parameter set from manual calibration of setup (1)

Performance criteria	Setup (1)		Setup (2)	
	Calibration	Validation	Calibration	Validation
NSE_{Qdaily}	0.59	0.60	0.72	0.71
r_{Qdaily}	0.79	0.80	0.87	0.87

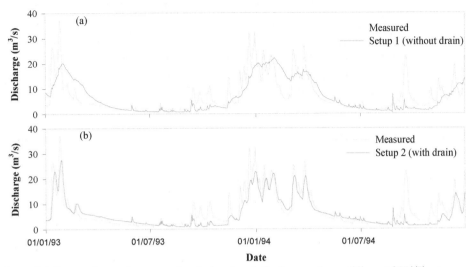

Figure 5.6 Comparison of measured and simulated discharges from different SWAT setups
using manual calibration at the gauging station 45_26

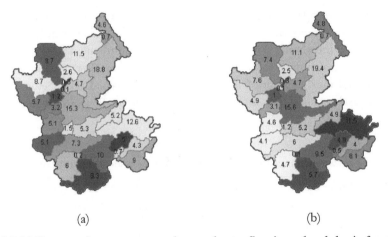

(a)   (b)

Figure 5.7 (a) Decrease in average annual groundwater flow in each sub-basin from setup (1)
to setup (2); and (b) Increase in average annual tile flow in each sub-basin from setup (1) to
setup (2)

5.3.3.2 Model performance in different structures with the help of auto-calibration

The two setups were calibrated by auto-calibration tool of SWAT at the station 45_26. The
most sensitive parameters were taken into account in the auto-calibration (see table 4.3 in
chapter 4). Table 5.6 presents the value of model performance criteria in daily time step of
the two setups and figure 5.8 illustrates the comparison of daily discharge values at station
45_26 of the two setups versus observations. The results show that with the help of auto-

calibration both models gave very good fit to measurements ($NSE_Q = 0.82 - 0.86$ in the calibration period). Setup (1) which does not include tile drain simulation gave better fit than setup (2) in which tile drainage parameters are considered. The result at station 45_21 and 45_01 which locate upstream of the calibrated station 45_26 were also checked. The same result was obtained for these two stations stating that setup (1) got a better fit than setup (2). The NSE values for these two stations were not as good as the calibrated station, however, the results were considered good for both setups.

Table 5.6 Values of daily NSE$_Q$ at the gauging stations for different SWAT setups after autocalibration

Station	Setup (1)		Setup (2)	
	Calibration	Validation	Calibration	Validation
45_26 (calibrated station)	0.86	0.84	0.82	0.80
45_21	0.85	0.80	0.75	0.75
45_01	0.84	0.74	0.83	0.77

Looking at the water balance in two setups (table 5.7), we can see the difference in the contribution of flow components. In setup (1), surface runoff was dominant flow component while tile flow and groundwater flow were both dominant contributions in setup (2). Because of high contribution of surface runoff which has shorter lag time, flow result in setup (1) was more dynamic and thus easily fitted to the variation of measured flow. Setup (2) has less dynamic flow result because they are dominated by subsurface flows which have longer lag time.

Table 5.7 Annual water balance for different SWAT setups after auto-calibration (1993-1998)

Water balance components	Setup (1) (mm)	Setup (2) (mm)
Precipitation	852	852
Surface runoff	249	52
Lateral soil	1	9
Tile flow	0	156
Groundwater flow	117	160
Evapotranspiration	474	461

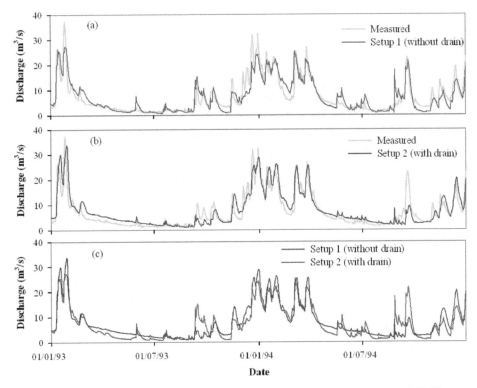

Figure 5.8 Comparison of measured and simulated discharges from different SWAT setups using auto-calibration at the gauging station 45_26

5.4 DISCUSSION AND CONCLUSIONS

5.4.1 Performance of different model structures in modelling flow and nitrogen fluxes

According to the model evaluation guidelines proposed by Moriasi et al. (2007) and statistical performance criteria established for the Danish national hydrological model (Henriksen et al., 2003), the flow results in the SWAT model were judged to be "very good" or "excellent". It can be confidently stated that SWAT accurately simulates the river basin discharge based on the statistical results and graphical comparison. However, literature information, field observations and expert knowledge all underscore that subsurface drainage is the dominant source of water to the Odense River stream network and that in turn very little surface runoff is observed. This indicates that SWAT may over-predict surface runoff and groundwater flow for the basin and in turn under-predict the tile drain contributions to streamflow. In contrast, the DAISY-MIKE SHE results reported by Hansen et al. (2009) present a more balanced flow representation with very low surface runoff and groundwater flow and high tile flow. The predicted nitrate flux pathways also varied considerably between the SWAT results reported here and the previous DAISY-MIKE SHE results (table 5.4), although the total simulated nitrate fluxes were consistent with measured fluxes in both studies based on the calibration results.

The comparison of different SWAT models also raised another issue. Although tile drains are considered the most important contribution to streamflow and the inclusion of tile drainage simulation is also verified to be considerably important to streamflow simulation, the SWAT setup (1) without tile drainage simulation achieved the best fit between the two models in terms of Nash-Sutcliffe coefficient with the help of auto-calibration. If only based on NSE, it can be stated that the SWAT setup (1) without tile drainage in which surface runoff is the dominant source for streamflow is better than the setup (2) with tile drainage because of the closer goodness-of-fit reflected by the higher NSE. However, it is known that most of the case study area is drained by tiles; tile flow is the most important source that contributes water to the streams and very little surface runoff is observed. Based on this information, the SWAT model without tile drainage in which surface runoff is the primary flow component is considered unrealistic due to neglecting the importance of tile drainage in the system.

From the above discussion it can be concluded that the decision which model is most appropriate to use must be based on knowledge of the dominating local processes and field measurements. For example, Vagstad et al. (2009) showed that seven different models that performed similarly in calibration against historical data had very different predictions for the effects of changes in agricultural practices, in a comparative study of nutrient leaching from agriculture. Similarly, Troldborg (2004) showed that among 10 hydrological models that were identical, except for differences in geological conceptualizations, the models that performed best during calibration against groundwater head and discharge data were not necessarily the ones that performed better during a subsequent solute transport prediction of environmental tracers in groundwater. These results supports the view expressed by Refsgaard and Henriksen (2004) and Stisen et al. (2011) that specification of multiple performance criteria reflecting the purposes of the specific modeling tasks is crucial for the ultimate success of a modeling study.

An important lesson learned in this study comparing different models that the structure of a specific model is very important in deciding whether the specific model is able to reflect targeted hydrologic and nutrient cycling conditions realistically. Different model structures result in different estimates in the water pathways to surface water, and also different pollutant loads to the streams. This was confirmed by Schoumans et al. (2009), who evaluated differences between eight model applications to quantify diffuse annual nutrient losses from agricultural land in three different river basins. They found that water and nutrient loads from river basins could be predicted, but that the nutrient pathways within agricultural land and nutrient losses to surface waters from agricultural land differed considerably between basins and model approaches. It is also clear that statistical metrics, which are important for evaluating the fitness of simulated results to measured data, are nevertheless unable to fully distinguish if a model is capable of illustrating water and pollutant pathways correctly. Therefore, it is extremely important to select a suitable model structure for a particular case study based on as much as possible information about the area, expert knowledge, observations in the field, etc.

5.4.2 Performance of tile drainage modelling in SWAT and DAISY-MIKE SHE

The Odense River basin is known as an intensely tile-drained basin; thus, tile flow is the major pathway of water to the stream system. The comparison between different SWAT models also verified that the inclusion of tile drainage simulation is very important to flow simulation and considerably improved streamflow results. The results showed that SWAT generated tile flow in high flow periods but not in low flow periods while DAISY-MIKE SHE was able to account for tile flow during low flow periods. It is noted that no direct field observations or measurements of tile flow in the low flow periods has been conducted in the Odense River basin region. Therefore, it cannot be concluded with absolute certainty that DAISY-MIKE SHE simulated the low flow periods accurately. However, it is clear that the two models responded very differently during the low flow periods, and it is likely that the DAISY-MIKE SHE provided a more accurate accounting of the actual hydrologic system.

DAISY-MIKE SHE is a comprehensive physics-based model that applies the 3D Boussinesq equation for saturated flow and requires an extensive amount of geological data. In DAISY-MIKE SHE, groundwater is routed from grids to grids, so it is possible for tiles in low areas to receive groundwater from the grids in upland areas. Therefore, in the low flow period, tile flow is still present which originates from groundwater from upland or neighboring areas.

In contrast, SWAT2005 simulates all hydrological processes at HRU level (Neitsch et al., 2005) and then sums runoff from each HRU to obtain the sub-basin water yield. There is no reference between the position of an HRU to landscape location and there is no routing of flow between HRUs. Thus, this approach fails to capture the interaction between the HRUs because they are not internally linked within the landscape but are routed individually to the basin outlet. Therefore, the effect of an upslope HRU on a downslope HRU cannot be evaluated (Arnold et al., 2010). Based on the HRU approach, in SWAT, tile flow in each HRU can only be generated with the water available in that HRU because there is no groundwater routing between HRUs or between sub-basins that makes it possible for tile drains to receive groundwater from neighboring areas. Therefore, it is possible that there was no water left to contribute to the tile flow during summer periods that were characterised by less rain water and higher evapotranspiration, which is probably the reason why SWAT cannot generate tile flow in the low flow period in this study.

It cannot be currently concluded as to which model definitely provided the best estimates of tile drainage for the Odense River basin region, due to the lack of tile drainage measurements. It may be argued that the most physically-based model should perform better because it best represents physically-based processes and properties; however, this is not a guarantee in itself. For example, Nasr et al. (2007) reported that the Hydrological Simulation Program – FORTRAN (HSPF) model performed better than the physically-based model SHETRAN. Meselhe et al. (2009) further found that the conceptual model HEC-HMS and physically based model MIKE SHE provided similar runoff prediction accuracy when data was available for calibration. The combination of physics-based small scale equations with detailed distributed modeling, may lead to equifinality and high predictive uncertainty (Savenije, 2010). This is because physics-based models are sometimes over-parameterized, and therefore, many different parameter sets will give almost identical fits to the measured data (the equifinality problem) but can yield dramatically different predictions of how the system will behave as conditions change. By making it easier for the model to get the right answer, over-parameterization makes it harder to tell whether they are getting the right answer for the right reason (Kirchner, 2006). Therefore, observations must be available for a wide range of conditions in order to choose the most absolutely accurate and realistic model possible.

The hydrological world is complex and heterogeneous and interactions always occur between different components of hydrological cycle. Therefore, the lack of interaction between HRUs absolutely needs to be improved in further development of SWAT, and this has been under development based on the hillslope approach introduced by Arnold et al. (2010). Arnold et al. (2010) introduced hillslope approach for SWAT modifications by applying and comparing four landscape delineation methods in SWAT: lumped, HRU, catena and grids in a small catchment to test the performance of different delineation. It was suggested that the catena or hillslope approach in which the basin was divided into three landscape units: the divide, hillslope and valley bottom and the routing between these three units is possible, may be a good alternative for large-scale applications because it preserves landscape position and allows riparian and flood plain areas to be simulated as discrete units. Therefore, development of functionalities to integrate the existing SWAT processes with hillslope processes appears to be an obvious way to enable SWAT to provide more realistic simulations of certain hydrological processes at both landscape and river basin scales. A simplified modification for SWAT following the hillslope approach was developed in this thesis and is described in the next chapter.

Chapter 6

THE APPROACH TO REPRESENT THE LANDSCAPE VARIABILITY IN THE SWAT MODEL

6.1   INTRODUCTION

The SWAT model simulates hydrological processes of a river basin by dividing it into multiple sub-basins, each of which is then divided into multiple Hydrological Response Units (HRUs). A HRU is a unique combination of soil and land use and slope in a single sub-basin. The flow result of a sub-basin is an aggregation of different types of flow generated from HRUs that are located inside the sub-basin. There is no reference between the position of an HRU to landscape location and there is no flow routing process between HRUs. Therefore, this approach fails to represent the interaction between an upland HRU and a lowland HRU (Arnold et al., 2010). The position and the connectivity of the different landscape elements may have a determining influence on the retention and transformation of many pollutants such as nitrogen.

In this chapter, we present a modification of the SWAT model (SWAT_LS) which accounts for the landscape position of HRUs and the routing of water and nitrogen across different landscape elements. First, we included the landscape variability in the HRU division step in order to define which landscape an HRU belongs to. Then, a flow and nitrate routing processes is added to route water and nitrate load from the upland landscape unit to the lowland landscape unit. The application of the SWAT_LS model is illustrated in a simple hypothetical case study which covers two landscape units: upland and lowland. Additionally, we carried out a sensitivity analysis on flow simulation using SWAT_LS and compared the results with the original SWAT model.

6.2   THE APPROACH TO REPRESENT THE LANDSCAPE VARIABILITY IN THE SWAT MODEL (SWAT_LS)

The approach to represent the landscape variability in SWAT (SWAT_LS) is described below by comparing with the original SWAT model. There are two main differences between the two approaches: (i) *HRU division* and (ii) *hydrological routing concept through different landscape units*. In this study, the version SWAT2005 was used.

6.2.1 HRU division

The new approach aims at accounting for landscape position and processes in the simulation of flow within sub-basin level. Figure 6.1 illustrates the division of a hypothetical sub-basin into HRUs in the original SWAT2005 and SWAT_LS. Originally, a HRU in SWAT is a unique combination of soil, land use and slope. In this simple illustration, we assume that the hypothetical sub-basin has homogenous slope, therefore, HRU generation is only based on soil and land use. In SWAT_LS, the division of sub-basin into HRUs is based on the overlay of soil, land use and landscape maps instead of the combination of only soil and land use map in the original SWAT2005 (figure 6.1). Therefore, the number of HRUs within one sub-basin will increase in the SWAT_LS. It is noted that figure 6.1 only illustrates an example of the division of a single sub-basin into HRUs while a real case study contains a number of such sub-basins.

The landscape map is created by dividing each sub-basin into several landscape units (LUs) which have different hydrological processes and transport mechanisms. In this study, the landscape map is simplified to only contain two LUs: upland and lowland. However, this approach can be applied for more LUs.

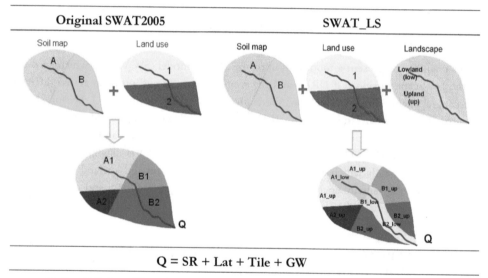

$$Q = SR + Lat + Tile + GW$$

Q: Overall discharge, SR: surface runoff, Lat: lateral flow, Tile: tile drainage, GW: groundwater flow

Figure 6.1 Comparison in HRUs division between the original SWAT and SWAT_LS

6.2.2 Hydrological routing concept through different landscape units

The discharge at the outlet of a sub-basin is the summation of four different flow components: surface runoff, lateral flow, tile flow and groundwater flow. In this modified approach, the hydrological routing from upland to lowland LUs is represented separately for each flow component in a simplified manner. Figure 6.2 describes the difference in hydrological routing between HRUs to the streams in two SWAT approaches. In the original SWAT2005, HRUs are individually routed directly to the river; therefore, the discharge to the river is the aggregation of flow generated from all HRUs. In SWAT_LS, there is routing from upland HRUs to lowland HRUs before reaching the river. The routing concept is modified in SWAT based on the routing concept of surface and subsurface lateral flow from terrain components introduced by Güntner and Bronstert (2004).

Surface runoff

In the original SWAT model, surface runoff at the outlet of a sub-basin is the aggregation of surface runoff generated from all HRUs located in the sub-basin. The SWAT_LS includes the interaction between HRUs in upland LU and HRUs in lowland LU by adding a routing concept from upland to lowland areas.

Surface runoff generated in the upland area (SR_{up}) is separated into (i) flow entering lowland component as runoff that is available for re-infiltration (SR_{up_low}) and (ii) remaining flow that goes directly to the river (SR_{up_direct}). The percentages of these two surface runoff components are assumed to be proportional to the respective areal fractions of landscape units. A larger lowland unit is assumed to be able to retain a larger fraction of runoff originating from the upland unit than a smaller one. This assumption is supported by the study of Güntner and Bronstert (2004) who used the same assumption to simulate the interaction of surface and subsurface lateral flow components from upslope topographic zones with those at downslope position. SR_{up_low} will be then added to the precipitation input of the lowland area to calculate surface runoff originating from lowland area (SR_{low}), infiltration and other processes. Lowland landscape unit is the last unit of the flow path; therefore, all surface runoff from lowland unit will go directly to the river.

In the example of figure 6.2, upland and lowland landscape units cover 70% and 30% of the area, respectively. Consequently, SR_{up_low} which accounts for 30% of surface runoff from upland (SR_{up}) can be retained by the lowland unit in which infiltration is allowed while the remaining SR_{up_direct} (70%) goes directly to the river. The total amount of surface runoff generated from the lowland area (SR_{low}) also flows to the river. Therefore, the total surface runoff from the sub-basin to the river (SR_{river}) is the summation of surface runoff from lowland (SR_{low}) and part of surface runoff from upland (SR_{up_direct}).

Lateral flow

Different from surface runoff, the total amount of lateral flow generated from the upland unit (Lat_{up}) goes to the lowland unit because this type of flow is a subsurface flow. In the lowland unit, Lat_{up} is considered as an additional contribution to the hydrological input of the lowland unit which is used to calculate lowland lateral flow (Lat_{low}) and other processes. The lateral flow that reaches the river equals to lateral flow from lowland unit.

Tile flow

Flow originating from tiles in the upland LU is assumed to go directly to tile storage in the lowland LU and join the lowland tile flow to the river. Tile flow from upland unit is distributed to HRUs in lowland unit based on the areas of HRUs in lowland areas. However,

HRUs in lowland areas may or may not have tile drains applied. Therefore, there are two cases to be considered in this tile flow routing:

- For an HRU in lowland areas which has tile drainage applied, tile flow received from upland area of this HRU is stored in its tile storage and joins its own generated tile flow to the river.

- For an HRU in lowland areas which does not have tile drain applied, tile flow received from upland area is considered as an additional hydrological input of the lowland unit, similar to lateral flow.

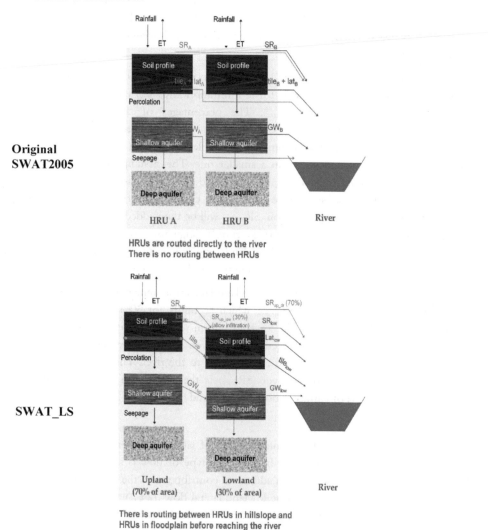

Figure 6.2 Difference in hydrological routing between HRUs to the river in two SWAT approaches

Groundwater flow

Groundwater generated from shallow aquifer in the upland unit is added to hydrological input of the shallow aquifer in the lowland unit. Then, groundwater from lowland LU is calculated and this is also the groundwater that flows to the river.

6.2.3 Methodology of applying the landscape concept in SWAT2005

To include the effect of landscape division in the hydrological modelling in SWAT, the landscape characteristic must be taken into account in HRU level. The procedure to simulate hydrological processes then also changes. While the original SWAT model simulates flow processes for all HRUs in a sub-basin and aggregate the results to get the flow value at the outlet of the sub-basin, the SWAT_LS calculates flow for HRUs in landscape units sequentially from the upland to lowland units and consider the flow routing between them.

Figure 6.3 shows the flowchart of steps to calculate hydrological processes in the SWAT_LS. There are 8 steps described as follows:

1. Define landscape IDs to all HRUs in sub-basins: This step aims at identifying which landscape that a HRU belongs to (upland or lowland unit).

2. Calculate processes for HRUs in the upland unit: SWAT_LS calculates all the hydrological processes in the HRUs of which landscape IDs correspond to upland.

3. Sum results of HRUs in the upland: The results of all HRUs in the upland are summed. These results are added as inflow to the subsequent LU, i.e. lowland.

4. Rout flow from upland to lowland: in this step, the routing concept as described above is applied to add flow from the upland as inflow to the lowland area for each HRU in lowland. For each flow component, the additional inflow from the upland unit is distributed to HRUs in lowland unit based on their areal fractions and then added to the own hydrological input of lowland HRUs.

5. Calculate processes in the lowland LU with additional input from the upland LU: In this step, SWAT_LS calculates hydrological processes for HRUs in the lowland unit with the new input which includes the additional inflow from the upland.

6. Sum results of HRUs in the lowland LU: the simulated results from HRUs in the lowland LUs are aggregated and the result of the outlet of the sub-basin will be calculated.

7. Write output results: the result at the outlet of the sub-basin is written in files for users. The results can be written in daily or monthly as required by users.

8. Check the water balance: The water balance at HRU level and sub-basin level is checked to eliminate mistakes in modelling with the modified approach.

In this thesis, we are also concerned about nitrogen simulation. With the adding of landscape routing processes, nitrogen processes were also modified similarly to flow. Nitrogen fluxes are first calculated for HRUs in upland LU. Due to its high mobility, only nitrate is routed from upland to lowland areas and participate in nitrogen processes in lowland areas before releasing to the river.

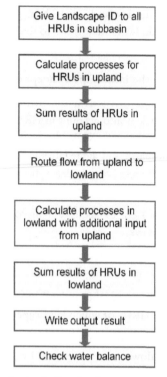

Figure 6.3 Methodology for including landscape processes in the SWAT model

6.3 TESTING THE SWAT_LS MODEL WITH A HYPOTHETICAL CASE STUDY

The existing code of SWAT version 2005 was used to modify to produce SWAT_LS. The SWAT_LS model was tested with a hypothetical case study. This is a simplest case study which only contains one HRU in each landscape unit. The objective of this test is to assess the performance of the modified approach as well as analyze the sensitivity of parameters in flow simulation.

6.3.1 Description

The hypothetical case study is a very simple case study with simple input data (figure 6.4). The case study has homogenous soil type, land use and slope. The area is divided into two landscape units: upland and lowland. Therefore, in this simple case, only two HRUs were created. Upland and lowland landscape units both include only 1 HRU. The meteorological data was taken from a single station which is located inside the case study. This is the simplest case study to test the routing concept between different landscape units. With a single HRU in each landscape unit, step 3 and 6 (figure 6.3) relating to the summations of results from HRUs in landscape units were ignored. Testing the new concept with such a simple case study is easier to realise and correct any possible mistakes in the calculation and understand the hydrological response of the modified concept.

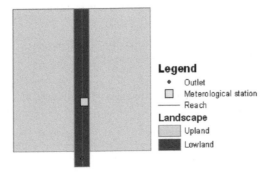

Figure 6.4 Hypothetical case study to test the SWAT_LS model

6.3.2 Sensitivity analysis of flow-related parameters in SWAT_LS in comparison with SWAT2005

Based on the sensitivity result from the set up of Odense river basin (see chapter 4), the parameters related to hydrological processes in the basin, not related to channel routing were chosen for this analysis. These parameters are listed in table 6.1.

To evaluate the effect of a parameter on the flow response, its value was adjusted inside the value range while the other parameters were kept at their default values. With each parameter sets, the model was run to obtain the flow results. For each parameter, the flow responses relative to its change inside the range were plotted in figure 6.5 for both SWAT_LS and SWAT2005. It is noted that the sensitivity of parameter 1 - 6 (table 6.1) were studied in a non-tile-drained condition while parameter 7 - 9 which are related to tile drain simulation were evaluated in a tile-drained situation.

Table 6.1 Most sensitive flow-related parameters, value ranges and default values

No.	Parameters	Description	Process	Default value	Range
1	*esco*	Soil evaporation compensation factor (-)	Evapo-transpiration	0.95	0 - 1
2	*epco*	Plant water uptake compensation factor (-)	Evapo-transpiration	1.0	0 - 1
3	*cn2**	SCS runoff curve number for moisture condition II	Surface runoff	0	-50 - 50 (%)
4	*surlag*	Surface runoff lag time (days)	Surface runoff	4	1 - 20
5	*alpha_bf*	Baseflow alpha factor (days)	Groundwater flow	0.048	0 - 1

No.	Parameters	Description	Process	Default value	Range
6	*sol_awc**	Available water capacity of the soil layer (mm/mm)	Percolation	0	-25 - 25 (%)
7	*dep_imp*	Depth to impervious layer for modeling perched water tables (mm)	Tile drainage	6000	0 - 6000
8	*tdrain*	Time to drain soil to field capacity (hrs)	Tile drainage	0	24-72
9	*gdrain*	Drain tile lag time (hrs)	Tile drainage	0	24-72

: parameter that is changed by percentage of the initial values

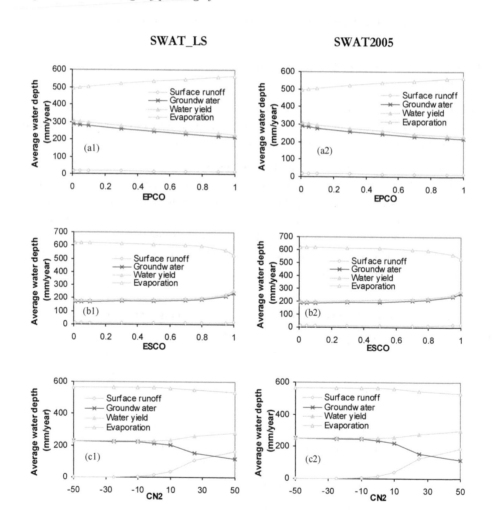

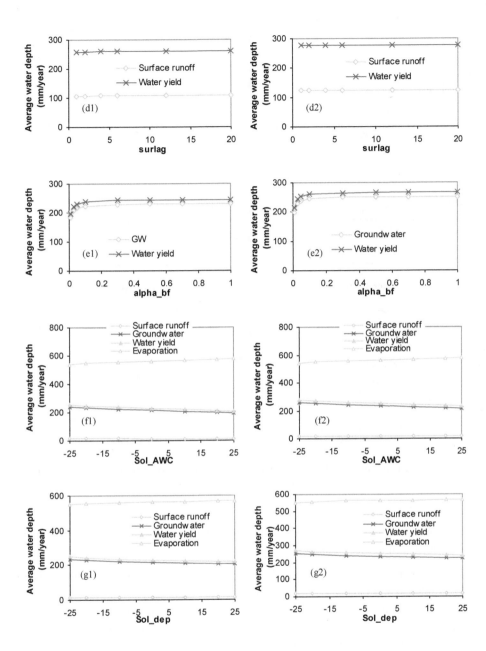

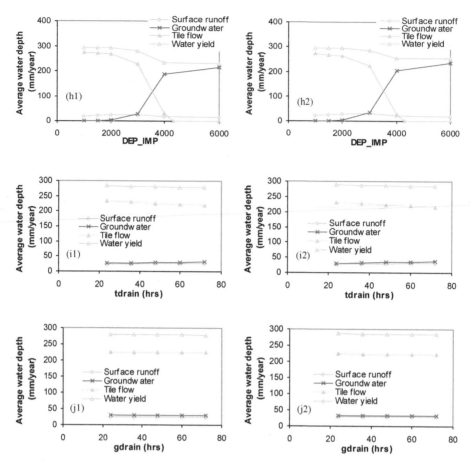

Figure 6.5 The sensitivity of flow-related parameters relative to annual flow response in SWAT_LS and original SWAT2005

The variation of flow response versus the change of parameters in the two approaches (figure 6.5) results in several issues to be discussed:

- The sensitivity of flow parameters in SWAT_LS is similar to their behaviours in SWAT2005. When parameters were changed inside the ranges, the flow magnitudes of the two approaches were slightly different, but the sensitivity trend is similar. SWAT_LS adds the division of sub-basin into two LUs (upland and lowland) and allows routing between them. The similarity in parameter sensitivity of the two models can bring to the conclusion that the added routing between two landscape units can affect the water volume but does not influence the flow behaviours in the sub-basin. This result is reasonable because our SWAT modification only aims at including landscape variability in the hydrological simulation keeping the same hydrological concepts and thus, keeping the same flow behaviour of the simulated case study.

- When the tile drain component is not considered, *cn2* is the most sensitive parameter. This parameter is the initial value for SCS runoff curve number which is then updated based on soil moisture or different management practice (plant, tillage or harvest/kill operation). In this test, the initial *cn2* equaled to 69, and the sensitivity of *cn2* was tested when *cn2* was changed from -50% to 50%. Figure 6.5 shows that surface runoff, groundwater flow and water yield were affected when the change of *cn2* is more than -10% (i.e *cn2* values > 60). When *cn2* value was smaller than 60, retention capacity of the basin which is related to parameter *cn2* was high, and no surface runoff was generated (figure 6.5). When *cn2* was higher than 60, the more *cn2* increased, a higher amount of surface runoff and lower amount of groundwater were generated. Although groundwater flow decreased, water yield tended to be higher relative to the increase of *cn2* and surface runoff because surface runoff is a fast flow component compared to groundwater flow. *Cn2* also affected evapotranspiration because the increase of surface runoff resulted in less water available for evapotranspiration.

- In tile-drain-applied condition, *dep_imp* is the most sensitive parameter which can change tile flow, groundwater flow and thus, water yield dramatically (figure 6.5h1 and 6.5h2). This parameter is to create an impervious layer in the model at which the water level rises and tile flow is generated if the water level is higher than the tile drain level. Figure 6.5h shows that tile flow only occurred when *dep_imp* was lower than 4.2m. When *dep_imp* was higher than 4.2m, the impervious layer was too deep that there was not enough water to rise above the tile drain level. When *dep_imp* was lower than 4.2m, tile flow ascended corresponding to the decrease of *dep_imp* because the water level was higher and thus more tile flow was generated. However, when *dep_imp* was close to the bottom of soil profile (1.5m in this case study) the amount of tile flow did not change much. The reason is that when *dep_imp* reached the bottom of the soil profile, all water infiltrated into the soil was kept inside the soil profile and thus, no groundwater was generated. This amount of water flowed to the river through tile flow or lateral flow. Compared to *dep_imp*, the other two parameters *tdrain* and *gdrain* which relates to the tile lag time have very little impact to annual amount of tile flow and other flow components (figure 6.5i and 6.5j).

- *Epco* and *esco* which are related to the simulation of evapotranspiration affected not only evaporation but also groundwater flow while having almost no effect on surface runoff. *Surlag* almost had no influence on annual surface runoff. *Alpha_bf* had higher impact on the decrease of annual groundwater flux and water yield when *alpha_bf* was lower than 0.1 while almost had no impact when it was higher than 0.1. The soil parameters *sol_awc* and *sol_z* slightly increased evaporation and decreased groundwater flow when their values were increased; however, the differences were not significant.

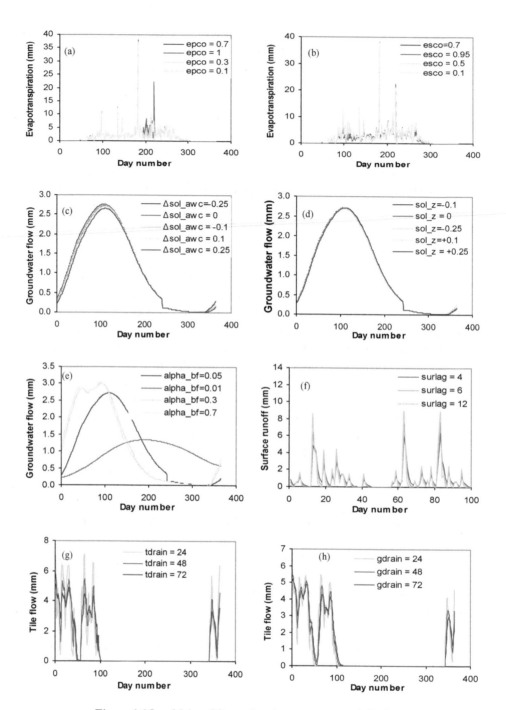

Figure 6.6 Sensitivity of flow-related parameters on daily flow

- From figure 6.5 we evaluated the sensitivity of parameters versus annual water fluxes. It is possible that one parameter is not sensitive for annual results, but can be very sensitive for daily results. *Surlag, alpha_bf, tdrain* and *gdrain* are parameters that are related to lag time of different flow components. They are not sensitive to the annual water flux results, but they would be the controllers of the flow variation on a daily time scale. Figure 6.6g and 6.6h shows that *tdrain* and *gdrain* relating to tile flow lag time did play an important role on the daily tile flow variation, i.e. the decrease of *tdrain* and *gdrain* gave higher peak flow and more dynamic fluctuation although they almost had no effect on annual water fluxes. *Alpha_bf* also had an effect on the delay of groundwater flow (figure 6.6e). *Surlag* also influenced surface runoff variation by increasing the peak flow as well as the sharpness of the recession curve when *surlag* ascended (figure 6.6f). *Esco* and *epco* had effect on evapotranspiration in some periods that the evapotranspiration demand needed to be met by water from the lower soil levels (figure 6.6a and 6.6b). At daily time scale, *sol_awc* which is the difference between soil field capacity and wilting point showed higher effect on groundwater flow in the first 4 months and the last 2 months of the year (winter) compared to the remaining months (summer) (figure 6.6c). The reason is that the increase of *sol_awc* in the condition of low temperature in winter allowed more water staying in the soil profile and less water becoming groundwater flow. On the other hand, high temperature in summer resulted in high evaporation which took more water in the soil and soil moisture was usually less than field capacity, thus the increase of *sol_awc* did not affect much flow results in this period.

6.3.3 Evaluation of the effect of parameter changes on the flow difference between SWAT_LS and the original SWAT2005

Figure 6.7 illustrates the difference between SWAT_LS and the original SWAT models for each water balance component in the context of parameter change. Water balance components include surface runoff, lateral flow, groundwater flow, tile flow, evapotranspiration. However, lateral flow is usually very small and thus, is ignored in this analysis.

- Generally, if the parameters in the two models are set at similar values, with any change of parameters, the added routing between landscape units in SWAT_LS decreases both surface runoff and groundwater flow but slightly increases tile flow compared to the original SWAT model in which HRUs are routed individually to the outlet of the sub-basin. The routing between landscape units allows a part of surface runoff from upland infiltrating back to the soil in lowland areas, thus surface runoff usually decreases if the floodplain soil is not saturated, and then more water reaches the shallow aquifer. Although there is more water reaching the shallow aquifer, the groundwater flow which is a slow component flowing from upland to lowland to reach the river (SWAT_LS) takes relatively more time than individual routing of groundwater in HRUs to the river (original SWAT), so the groundwater flow decreases in SWAT_LS. Different from groundwater flow, more water reaching the shallow aquifer results in more tile flow reaching the river in SWAT_LS. The reason is that tile flow is considered a fast flow component which does not take long time to reach the river, therefore, the rise of water in shallow aquifer results in the increase of tile flow. The added routing between landscape units rarely change annual evapotranspiration result if the amount of water in the system is kept the same.

- The difference in flow components between two models was most susceptible to *cn2* and *dep_imp*. This result is easily explained because *cn2* and *dep_imp* are also the most sensitive parameters to flow response as discussed above; therefore, it is reasonable that they also affected the flow differences the most.

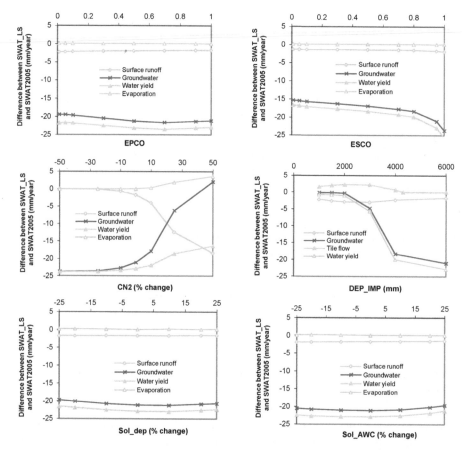

Figure 6.7 Sensitivity of flow-related parameters relative to the flow difference between SWAT_LS and SWAT2005

- When there is tile drain installed in the case study, flow difference between two models is also sensitive to *dep_imp*. When *dep_imp* was lower than 4.2m, tile flow started to be generated. Tile flow difference between two SWAT models was positive which mean that SWAT_LS generated more tile flow than the original SWAT. Tile drain helps to decrease soil moisture in the soil, thus, also affect surface runoff simulation.

Groundwater difference was the variable that had greatest effect from *dep_imp*. When *dep_imp* was low enough for tile flow to occur, simulated groundwater as well as the groundwater difference between two models decreased. When *dep_imp* was lower than 2 m, groundwater was rarely generated (figure 6.5h), and therefore, this difference kept the value at around zero.

- The change of parameters related to evapotranspiration: *epco*, *esco* did not lead to any evapotranspiration difference between the two models. The flow difference was not sensitive to *epco* for all three flow components while *esco* did have only little impact on the groundwater difference. According to the parameters related to soil characteristics *sol_awc* and *sol_z*, the flow did not change when these parameters were adjusted, because both models were not very sensitive to these parameters as discussed above.

6.3.4 The effect of areal proportion of landscape units on the response of flow components

In the previous section, the sensitivity of parameters and flow variation were discussed in a case study in which lowland covers 12% of the area. In this section, the effect of areal proportion of two landscape units, i.e. upland and lowland, on the response of flow components is discussed. Table 6.2 summarizes the annual flow components and figure 6.8 shows the flow variation simulated in the year 1994 of three flow components: surface runoff, groundwater and tile flow in different scenarios in which the lowland covers different percentages of the area. Graphs on the left present the flow generated from upland areas while the ones on the right illustrate the flows from the whole sub-basin that reach the river.

Table 6.2 Annual flow components in two cases (i) no tile drain installed and (ii) tile drain installed with different percentages of lowland area (unit: water depth mm)

Scenarios	No tile drain installed				Tile drain installed			
	Surface runoff (mm)		Groundwater flow (mm)		Tile flow (mm)		Groundwater flow (mm)	
	From upland	Total flow	From upland	Total flow	From upland	Total flow	From upland	Total flow
Low_0	173.9	173.9	91.6	91.6	211.5	211.5	34.1	34.1
Low_20	139.6	158.9	72.9	107.4	169.2	219.0	27.1	35.2
Low_40	105.0	151.5	54.4	116.8	126.9	218.2	20.2	36.5
Low_50	87.6	150.4	45.2	118.3	105.7	217.2	16.8	37.1
Low_60	70.28	151.2	36.0	118.3	84.6	215.7	13.4	37.8
Low_80	35.2	159.2	17.9	114.0	42.3	211.7	6.7	39.1
Low_100	0.0	176.0	0.0	104.5	0.00	206.0	0.0	40.5

Note: Low_X means that the lowland area covers X percent of the sub-basin area.

Table 6.2 shows the flow results in 2 cases (i) no tile drain installed and (ii) tile drain installed. The results show that the areal proportion of upland and lowland units determined the amount of flows from the upland area. For all flow components, the expansion of the lowland area resulted in the decrease of flows from upland while it had a slighter effect on the total flow from the whole sub-basin.

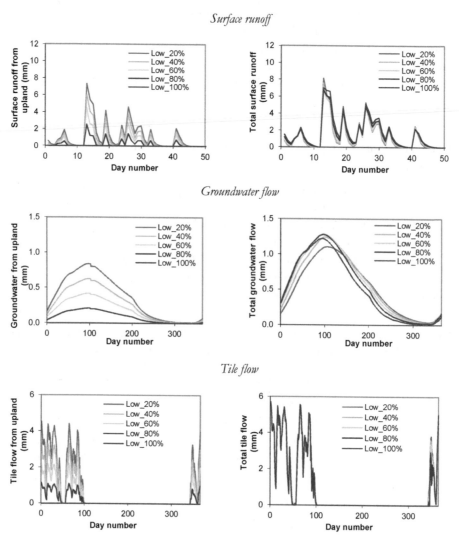

Figure 6.8 The effect of proportion distribution of landscape units on the flow from upland area and total flow from the whole sub-basin

- In case (i), surface runoff and groundwater were dominating flow. When the lowland covered less than 50% of the sub-basin area, the expansion of the lowland area resulted in a slight decrease of surface runoff correlated to a small increase of groundwater flow because the lowland area helped to retain surface runoff and then increased water contribution to the groundwater aquifer. On the other hand, when the lowland covered a larger area exceeding 50% of the sub-basin, surface runoff started to slightly increase, which corresponded to a small decrease of groundwater because flow in the lowland area went directly to the river with lower delay time. For daily surface runoff, the expansion of the lowland area decreased peak flow because of the delay of surface runoff. According to groundwater flow, the percentage of lowland area had a significant influence on delaying groundwater. The effect of groundwater delay is higher when the upland covered larger area because in this condition, a larger amount of groundwater from upland was delayed by the lowland area.

- In case (ii), tile flow and groundwater flow were considered as dominating flow. In this case, tile drains were assumed to be installed in both upland and lowland and helped to release excess water from upland via lowland to the river. From table 6.2 and figure 6.8, it can be clearly observed that the areal proportion of landscape units did not have a significant effect on tile flow both on annual and daily time scale. It is because tile flow is a fast flow component and its amount was not lost for any processes through its way to the river.

6.4   CONCLUSIONS

The approach chosen to represent the landscape variability in SWAT (SWAT_LS) is to divide a sub-basin into different landscape units and allow hydrological routing between them. For all flow-related parameters, their sensitivities to flow response in SWAT_LS are similar to their behaviours in the original SWAT2005 model. Therefore, it can be concluded that the added routing between landscape units can affect the water volume but does not influence the flow behaviour within the sub-basin. Curve number *cn2* and depth of impervious layer *dep_imp* are the most sensitive parameters not only to flow response in each model but also to flow differences between the two models, i.e. SWAT_LS and SWAT2005. Generally, compared to SWAT2005, SWAT_LS is seen to decrease both surface runoff and groundwater flow but slightly increase tile flow in case tile drains are applied in all landscape units. Moreover, the areal proportion between upland and lowland areas does seem to have a very strong effect on upland flows; however, the effect on flows from the whole sub-basin is not significant.

Chapter 7

INTEGRATING A CONCEPTUAL RIPARIAN ZONE MODEL IN THE SWAT MODEL

7.1 INTRODUCTION

This chapter describes the Riparian Nitrogen Model (RNM), a conceptual riparian zone model for simulating nitrate removal by denitrification, and the incorporation of this model into the SWAT model by modifying the Fortran codes of SWAT. Presently, SWAT does have a module to represent riparian zones/filter strips, but this module is limited to estimating the efficiency of flow and nutrient retention by empirical equations. The Riparian Nitrogen Model gives a better representation of nitrate removal by denitrification by simulating this process via two mechanisms: (i) groundwater passing through the riparian buffer before discharging into the stream and (ii) surface water being temporarily stored within the riparian soils during flood events, with the assumption that denitrification declines with depth because of the availability of organic matter. The modified SWAT model was then tested with a simple hypothetical case study in different scenarios.

7.2 DESCRIPTION OF THE RIPARIAN NITROGEN MODEL (RNM)

The Riparian Nitrogen Model (RNM) (Rassam et al., 2008) is a conceptual model that estimates the removal of nitrate as a result of denitrification, which is one of the major processes that leads to the permanent removal of nitrate from shallow groundwater during interaction with riparian soils. The denitrification occurs when groundwater and surface water interact with the riparian buffers. This interaction occurs via two mechanisms: (i) groundwater flow through the riparian buffer zone, and (ii) temporary surface water storage within the riparian soils during flooding. The model operates at two conceptual levels based on stream orders: ephemeral low order streams and perennial middle order streams. Ephemeral streams are conceptualised as being streams that do not receive any kind of permanent flow/interflow component, but rather are channels for quick flow during events. Perennial streams are conceptualised as being those streams that do receive a permanent base flow/ interflow component. In ephemeral low-order streams, a simple bucket model is used. Areas of potential groundwater perching are identified. During flood events, those areas fill like a bucket (surface water becomes groundwater), which is then denitrified during the flood event, and subsequently drains back to the surface water system. In perennial middle-order streams, denitrification occurs when base flow intercepts the rootzone. A shallow water table and a high residence time promote denitrification. Denitrification may also occur when stream water is temporarily stored in banks during flood events.

The amount of water stored in banks depends on the size of the flood event, the soil properties such as hydraulic conductivity and porosity, the geometry of the floodplain and the residence time. Figure 7.1 shows the various processes considered in the RNM and how denitrification is modelled.

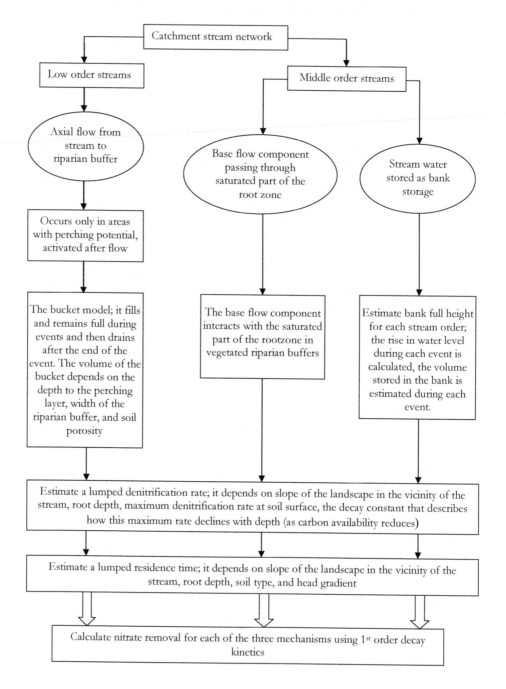

Figure 7.1 Flow chart showing various processes in the Riparian Nitrogen Model (Rassam et al., 2005)

7.2.1 Modelling variable denitrification rates through the soil profile

In the RNM, the distribution of denitrification potential with depth is modelled using an exponential decay function:

$$R_d = R_{max} \frac{e^{-kd} - e^{-kr}}{1 - e^{-kr}} \tag{7.1}$$

where d is the vertical depth below the ground surface (L; where L refers to length units), R_d is the nitrate decay rate at any depth d (T^{-1}; where T refers to time units), R_{max} is the maximum nitrate decay rate at the soil surface (T^{-1}), r is the depth of the root zone (L), and k is a parameter describing the rate at which the nitrate decay rate R declines with depth (L^{-1}).

Denitrification processes are assumed to be negligible below the rooting depth as insufficient carbon is available (Rassam et al., 2008). Therefore, equation 7.1 ensures zero denitrification below the rooting depth ($R_d = 0$ at $d \geq r$)

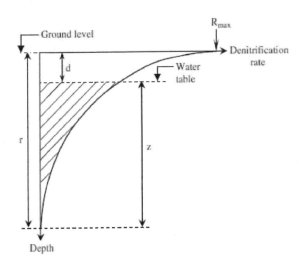

Figure 7.2 Distribution of denitrification rate through the riparian buffer (Rassam et al., 2005)

7.2.2 Conceptual models for potential denitrification

7.2.2.1 Model for ephemeral streams

In the low-order ephemeral streams, the stream water is likely to interact with the carbon-rich root zone of the riparian buffer in areas where a localised perched shallow groundwater table is formed. This happens when a low conductivity confining layer underlies the permeable soil of the floodplain. Denitrification in these riparian zones is likely to take place during flow events after surface water flows laterally from the stream to form a shallow perched water table in the floodplain and then later drains back to the stream.

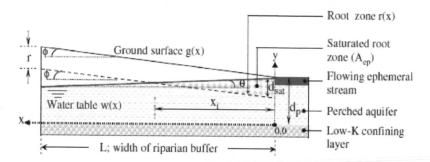

Figure 7.3 Conceptual model for denitrification in the floodplain of ephemeral streams (Rassam et al., 2008)

The length of the saturated root zone $d_{sat}(x)$ at any distance x is defined as:

$$d_{sat}(x) = w(x) - r(x) \tag{7.2}$$

where $w(x)$ water table at distance x and $r(x)$ is the root zone at distance x

$$w(x) = d_p - nx \tag{7.3}$$

$$r(x) = (d_p - r) + mx \tag{7.4}$$

where d_p is the depth to the low conductivity confining layer, $m = \tan(\phi)$ and $n = \tan(\theta)$ and ϕ

and θ are the inclinations of the ground surface and water table, respectively.

x_r, the active width of the riparian zone is defined as:

$$x_r = \min(x_i, L) \tag{7.5}$$

where L is the width of the vegetated riparian zone, x_i is the distance at which the water table intersects the root zone and is calculated as follows:

$$x_i = \frac{r}{m+n} \tag{7.6}$$

The area of the saturated root zone in which denitrification occurs is defined as follows

$$A_{ep} = \int_0^{x_r} d_{sat}(x)dx = rx_i - \frac{1}{2}x_r^2(m+n) \tag{7.7}$$

Since the denitrification rate is calculated based on the variable d which is the depth from ground surface (equation 7.1), the variable d is transformed in term of the coordinate system (x,y) as follows:

$$d(x, y) = d_p + mx - y \tag{7.8}$$

The denitrification rate at any depth $d(x,y)$ is calculated based on equation 7.6. Therefore, the average denitrification rate across the entire active section of the riparian buffer (A_{ep}) is

$$R_u = \frac{1}{A_{ep}} \int_0^{x_t} \int_{r(x)}^{w(x)} \frac{e^{-kd(x,y)} - e^{-kr}}{1 - e^{-kr}} \, dx dy \tag{7.9}$$

Integrating equation 7.9 we obtain

$$R_u = \frac{1}{A_{ep}} \frac{e^{-kr} R_{max}}{k(1-e^{-kr})} \left[\frac{1}{e^{-kr}(m+n)} (1 - e^{-kx_r(m+n)}) - x_r(1+kr) + \frac{kx_r^2}{2}(m+n) \right] \tag{7.10}$$

7.2.2.2 Models for perennial streams

In the case of perennial streams, the RNM simulates denitrification processes via two mechanisms: (i) groundwater passing through the riparian buffer (base flow) and (ii) surface water beijng temporarily stored within the riparian soils during flood event (bank storage).

✓ **Base flow model**

Figure 7.4 presents the conceptual model for floodplain denitrification occuring as groundwater flows laterally through the saturated part of the root zone. The water table is assumed to be a linear function of slope, which is equal to that of the ground surface and also the rootzone. Therefore, the groundwater table is parallel to ground surface and the bottom of the rootzone. The saturated part of the root zone extends across the entire width of the riparian zone (figure 7.4).

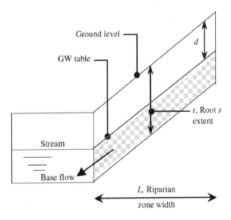

Figure 7.4 Conceptual model for floodplain denitrification, base flow component in perennial streams (Rassam et al., 2008)

As mentioned above, it is assumed that denitrification occurs only when the groundwater table is located inside the root zone. Therefore, the cross-sectional area exposed to denitrification is calculated for the entire width of the riparian zone as follows:

$$A_{bf} = L(r - d) \qquad for \quad w < r \tag{7.11}$$

where r is the extent of the root zone from the soil surface, d is the groundwater depth from the soil surface, and L is the riparian zone width.

Denitrification rate is decreased along depth inside the root zone area based on equation 7.1. In this conceptual model, groundwater table is assumed to be parallel to the ground surface and the bottom of the root zone (figure 7.4), thus the depth of the saturated zone where denitrification processes may occur is the same along the width of riparian zone. This also means that denitrification rate at a certain depth is constant along the riparian zone. Therefore, denitrification rate only varies along depth. The average denitrification rate R_u across the saturated zone is calculated as follows:

$$R_u = \frac{R_{max}}{(r - d)} \int_d^r \frac{(e^{-ky} - e^{-kr})}{1 - e^{-kr}} dy \tag{7.12}$$

where y is the vertical depth below the ground surface (L). Note that $R_u = 0$ for $d \geq r$.

Integrating equation 7.12, we obtain

$$R_u = \frac{R_{max}}{(r - d)(1 - e^{-kr})} (\frac{e^{-kd} - e^{-kr}}{k} + (d - r)e^{-kr}) \tag{7.13}$$

✓ **Bank storage model**

Figure 7.5 shows the conceptualisation of denitrification caused by bank storage in a perennial stream. The concept described in this section is a simplified bank storage model which ignores the time lags associated with filling and draining of water during and after flood waves.

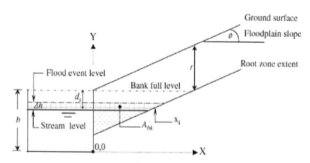

Figure 7.5 Conceptual model for denitrification during bank storage in perennial streams (Rassam et al., 2008)

As a flood wave passes, the water level in the stream increases by an amount of Δh, and saturates an area of the floodplain A_{bk}. The residence time can be expressed as the duration between the beginning of the flood event and the start of the recession period. During an event, water temporarily stored in the area A_{bk} loses nitrate through denitrification and as the flood recedes, it drains back to the stream with a lower nitrate concentration.

The saturated area A_{bk} on one side of the stream is:

$$A_{bk} = \frac{\Delta h}{2}(\min(\max(\frac{r-d_s}{\tan(\phi)},0),L) + \min(\max(\frac{r-d_s+\Delta h}{\tan(\phi)},0),L)) \tag{7.14}$$

where Δh is the change in the height of the stream during a flood event; d_s is the depth to stream water level, L is the width of the riparian zone, r is the depth of the root zone, ϕ is the slope of the floodplain.

Since the denitrification rate is calculated based on the variable d which is the depth from ground surface (equation 7.1), the variable d is transformed in term of the coordinate system (x,y) (figure 7.5) as follows:

$$d(x,y) = x\tan\phi + b - y \tag{7.15}$$

The average denitrification across the saturated area A_{bk} is calculated as follows:

$$R_u = \frac{R_{max}}{A_{bk}} \int_{b-d_s}^{b-d_s+\Delta h} \int_0^{x_i} \frac{(e^{-kx\tan\phi-kb+ky} - e^{-kr})}{1-e^{-kr}}\,dxdy \tag{7.16}$$

Integrating equation 7.16, we obtain

$$R_u = \frac{R_{max}}{A_{bk}(1-e^{-kr})}\left(\frac{(1-e^{k\Delta h})e^{-kd_s}(e^{-kx_i\tan\phi}-1))}{k^2\tan\phi} - \Delta hx_i e^{-kr}\right) \tag{7.17}$$

where x_i is the point where the water table intersect the root zone extent which is defined by:

$$x_i = \frac{r-d_s+\Delta h/2}{\tan\phi} \tag{7.18}$$

7.2.3 Reduction in nitrate loads caused by denitrification

The denitrification rates resulted from the above conceptual models are then used to calculate nitrate removal based on a simple 1st order decay function (equation 7.19).

$$C_t = C_0 e^{-R_u t} \tag{7.19}$$

where C_t is nitrate concentration at any time t (M/L^3); where M refers to mass units.

The suitability of the 1st order decay function in modelling the change of nitrate concentration by denitrification was test by Rassam et al. (2005) based on field observations.

7.3 ADDING THE RIPARIAN NITROGEN MODEL IN THE SWAT-LS MODEL

The SWAT_LS model simulates the hydrological and nitrogen processes in two different landscape units: upland and lowland and allows interactions between the two units. In order to model the riparian zone processes in the SWAT model, the lowland landscape unit is considered as riparian zone which receives flow and nitrate from upland areas and then releases them to the streams.

It is noted that the RNM only deals with denitrification process in riparian zones. In this study, the model was used to replace the current denitrification concept in the lowland/riparian zone unit. Other nitrogen related processes such as mineralization, immobilization, plant uptake that contribute or extract nitrate from the nitrate pool were kept unchanged.

As described in section 7.2, the RNM deals with denitrification in riparian buffers of ephemeral and perennial streams. The three conceptual models in the RNM, i.e. simple bucket model for ephemeral streams, base flow and simplified bank storage models for perennial streams are all based on the assumption that the denitrification rate declines with depth following an exponential equation as equation 7.1. For simplicity, to integrate the RNM in the SWAT model to simulate the denitrification process in riparian zones, we only consider the two concepts for perennial streams, i.e. base flow model and simple bank storage model. The red squares in figure 7.6 and 7.9 show where the RNM is applied in SWAT.

7.3.1 Applying the base flow model of RNM in the SWAT_LS model

The base flow model in the RNM is used to estimate nitrate removal by denitrification occurring in the floodplain/riparian LU when base flow passes through riparian zones and interacts with carbon-rich root zone here. Using the SWAT_LS model, we consider lowland LU as riparian zones. The RNM base flow model was used to replace the current denitrification process in the lowland LU.

To make it clear about the nitrate related process and where RNM is applied, figure 7.6 illustrates nitrogen transformation and transport processes in the SWAT_LS model. In upland LU, NO_3 in soil derives from rainfall, fertilizer application, mineralization and nitrification and NO_3 exists in the soil itself. NO_3 is lost by plant uptake and denitrification, and then follows the soil water to percolate into the shallow aquifer or leave to the next landscape unit, i.e. riparian zone, through surface runoff, tile flow and lateral flow. A part of NO_3 that leaches to shallow aquifer is also lost by biological/chemical processes which represents as a half-life nitrate parameter in SWAT. In lowland/riparian LU, NO_3 not only comes from rainfall, fertilizer, mineralization and nitrification, but also is added by NO_3 from upland LU through different flow components. NO_3 is also lost from plant uptake and denitrification and follows soil water to shallow aquifer or to the streams. The denitrification here (the red square in figure 7.6), in floodplain/riparian LU is performed by the RNM conceptual base flow model (as equation 7.13).

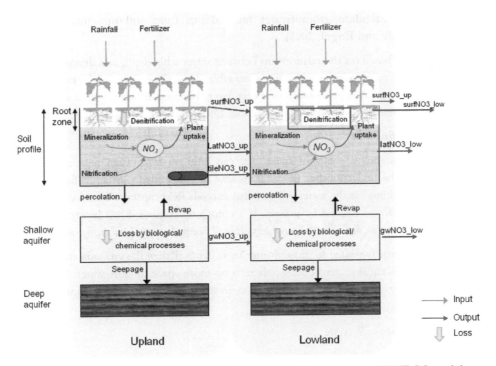

Figure 7.6 Nitrogen transformation and transport processes in the SWAT_LS model

From figure 7.4 and equation 7.13, it is observed that the groundwater table is a necessary variable to define the area over which the base flow interacts with the carbon-rich rootzone and thus is exposed to denitrification. Unfortunately, the original SWAT model does not calculate and provide groundwater table results. Vazques-Amabile and Engel (2005) proposed a procedure to compute perched groundwater depth using SWAT soil moisture outputs, based on the theory used by DRAINMOD, in order to expand SWAT's capabilities. The procedure was tested through calibration and validation for three sites located within the Muscatatuck River basin in Southeast Indiana. The results showed reasonable predictions for seasonal variation of groundwater table with correlation coefficients from 0.45-0.68 for three wells during the validation period. However, in the study of Vazques-Amabile and Engel (2005), the procedure to predict groundwater table from SWAT soil moisture results was implemented outside of SWAT model using a text file result from SWAT and without modifying the code of SWAT. In our study, because groundwater table is a necessary variable to calculate denitrification rate at every time step, we implemented the procedure proposed by Vazques-Amabile and Engel (2005) in the code of SWAT. The procedure to predict groundwater table from SWAT soil moisture results is described as follows.

❖ **Procedure to calculate groundwater table depth from soil moisture in SWAT (Vazquez-Amábile and Engel, 2005)**

This procedure is based on the relationship between water table depth and drainage volume, which is the effective air volume above the water table. This relationship can be calculated for every soil from the drainage volume of every layer, building the curve that depicts that relationship. The theory for this procedure follows DRAINMOD model (Skaggs, 1980).

Some related definitions

The "drainage volume" is the void space that can hold water between field capacity and saturation. It can be understood as the total volume of voids filled with air at field capacity. The "drainable volume" is the water volume that exceeds field capacity, is stored in the void space and is able to be drained by gravity. If a completely saturated soil is left to drain by gravity, the volume that drain, i.e. "drainable volume" is equal to the "drainage volume".

The drainage volume is related to groundwater table depth. Below the groundwater table, the drainage volume is equal to zero because there is no more space to store water. If the water table is lowered by ΔH (mm), the water drained, in terms of water depth (mm) will be equal to drainage porosity (S_y) multiplied by ΔH :

$$Drainage\ volume = S_y \times \Delta H \tag{7.20}$$

where drainage porosity S_y is computed as follows:

$$S_y = Porosity - Field\ capacity = Porosity - AWC - WP \tag{7.21}$$

where AWC is available water content which is a SWAT input, WP is wilting point and porosity are calculated by the SWAT model using the below equations.

$$Porosity = 1 - \left(\frac{bulk\ density}{2.65} \right) \times 100 \tag{7.22}$$

$$WP = 0.4 \times Clay(\%) \times \left(\frac{bulk\ density}{100} \right) \tag{7.23}$$

In a soil profile that has several layers, drainage porosity is calculated for each layer and drainage volume for each layer (unit: water depth) is calculated based on soil water stored in it. The total drainage volume for the whole soil profile is the integration of drainage volume in all the layers in the soil profile. The groundwater table depth is predicted from the total drainage volume of the soil profile based on a soil profile water yield curve which is built based on soil characteristics. The detailed procedure is described as below.

Steps to predict groundwater depth from soil water results in SWAT

Step 1: Build a soil profile water yield curve for each soil type

If a soil profile has several layers, the soil profile water yield curve, which shows the relationship of total drainage volume and water table depth, is calculated from layer drainage volume and the layer depth. It is assumed in each layer, the relationship between drainage volume and water depth is simplified as a linear function where the drainage porosity is the slope, i.e. the relationship follows equation 7.20.

To make it clear, we use a simple example from Vazquez-Amábile and Engel (2005) to build a soil profile water yield for a hypothetical two-layer soil (layer 1: 0-30cm, and layer 2: 30-70cm). The calculation of drainage volume for each layer and cumulative drainage volume are shown in table 7.1. The soil profile water yield curve is then plotted to show the relationship between cumulative drainage volume and water depth (figure 7.7).

Table 7.1 Calculation for building a soil profile water yield curve

(example from Vazquez-Amábile and Engel (2005)

Layer	Depth (cm)	Drainage porosity	Drainage volume $(S_y \times \Delta H)$	Cumulative drainage volume
Layer 1	0	0.12	0.12 * 0 = 0	0
	30		0.12 * 30 = 3.6	3.6
Layer 2	100	0.04	0.04 * 70 = 2.8	6.4

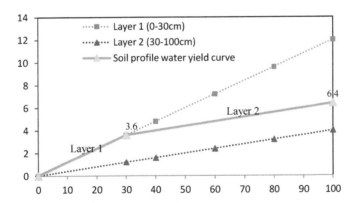

Figure 7.7 Layers and soil profile water yield curves (Skaggs, 1980)

If the soil profile has more layers, the soil profile water yield curve is as illustrated in figure 7.8

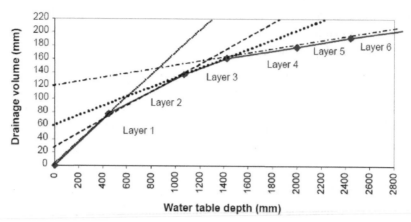

Figure 7.8 Soil profile water yield curve for a multi-layers soil

(Vazquez-Amábile and Engel, 2005)

<u>Step 2</u>: Calculate the total drainage volume for HRUs

SWAT computes daily soil moisture for every layer in the soil profile for every HRU in the simulated river basin. From soil moisture using the calculated soil profile water yield curve, water table depth is calculated as follows:

- Compute the amount of excess water above field capacity for every layer
- Calculate drainage volume for each layer (at the current groundwater table that needs to be found) by subtracting layer drainage volume at field capacity to the water excess above field capacity
- Summarise daily drainage volume for all layers to get the total drainage volume of the soil profile
- Intercept the value of total drainage volume of the soil profile to the soil profile water yield curve to obtain the daily groundwater table depth.

Limitations of the approach

The approach assumes a simplified linear relationship between soil drainage volume and water table depth. With this assumption, it means that the soil is completely drained immediately above the water table. However, there is a transition zone, or capillary fringe, above the water table. This transition zone is more important in fine soils than in coarse soil, which is not considered in this approach. Therefore, the soil profile water yield curve may differ from the actual soil characteristics and thus the results of water table depth may be different from reality. However, this error can be reduced through calibration.

7.3.2 Applying the simplified bank storage model of RNM in the SWAT-LS model

The bank storage model in the RNM deals with denitrification process when surface water interacts with riparian buffers during flood event. SWAT does simulate bank storage in the water routing process; however, no denitrification occurring in bank storage is taken into account. In this study, denitrification in bank storage is added in the SWAT-LS code based on the bank storage model of RNM (denitrification rate is estimate based on equation 7.17).

7.3.2.1 Current bank storage calculation in the SWAT model

Bank storage is calculated in the SWAT model as follows (Neitsch et al., 2005):

- Firstly, transmission losses through the side and bottom of the channel are estimated during periods when a stream receives no groundwater contributions.

$$tloss = K_{ch} \cdot TT \cdot P_{ch} L_{ch} \qquad (7.24)$$

where $tloss$ is channel transmission losses (m³), K_{ch} is the effective hydraulic conductivity of the channel alluvium (mm/hr), TT is the flow travel time (hr), P_{ch} is the wetted perimeter (m), and L_{ch} is the channel length (km).

- Transmission losses are assumed to enter bank storage or the deep aquifer. The amount of transmission losses entering bank storage (bnk_{in}) and the bank storage at each time step are calculated ($bnkstor$) as follows:

$$bnk_{in} = tloss(1 - fr_{trns}) \qquad (7.25)$$

$$bnkstor(i) = bnkstor(i-1) + bnk_{in} \qquad (7.26)$$

where fr_{trns} is the fraction of transmission losses partitioned to the deep aquifer, $bnkstor(i)$ is the bank storage on day i (m³), $bnkstor(i-1)$ is the bank storage on day $i-1$ (m³)

- Thirdly, bank flow (V_{bnk}) which is the flow that bank storage contributes to the reach is calculated with a recession curve similar to that used for groundwater simulation:

$$V_{bnk} = bnkstor(i) \cdot (1 - e^{-\alpha_{bnk}}) \qquad (7.27)$$

where a_{bnk} is the bank flow recession constant

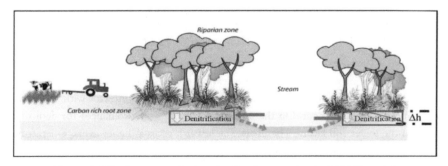

Figure 7.9 Bank storage during the flood events in the SWAT model

7.3.2.2 Adding denitrification process in bank storage to the SWAT model

The denitrification in bank storage was calculated through several steps:

- Based on the results of bank storage from SWAT, nitrate flux entering the bank storage is calculated based on the estimated nitrate concentration in the stream and flow entering bank storage at each time step.
- The depth of bank storage is calculated based on the bank storage volume and the area of riparian zone. Then the area where denitrification possibly occurs is estimated.
- Amount of denitrification is calculated assuming that denitrification rate declines through depth
- Nitrate flux that returns to the river from the bank storage as well as the new nitrate concentration in the stream are estimated

7.4 TESTING THE SWAT_LS MODEL FOR DENITRIFICATION IN RIPARIAN ZONES IN A HYPOTHETICAL CASE STUDY

7.4.1 Description of the hypothetical case study

The hypothetical case study used to test the added RNM in SWAT_LS is similar to the hypothetical case study mentioned in chapter 6. The case study has two HRUs each of which represents a landscape unit: upland and lowland. In this test we applied different vegetation for two landscape units: barley with fertilizer applied for the upland area which is considered as agricultural area, and grass for the lowland area without fertilizer application which is considered as riparian buffer zone. The root zone was assumed to be the same with the soil profile which has a depth of 1500 mm. Parameters in two landscape units were changed based on scenarios that we want to test the SWAT_LS model.

7.4.2 Sensitivity of parameters related to the simulation of denitrification in riparian zones

Following the equation 7.1, there are three parameters that affect the denitrification rate in the riparian zone:

- R_{max}: the maximum denitrification rate at the soil surface
- k: the rate at which the denitrification rate declines with depth
- r: the depth of root zone

Figure 7.10 to 7.12 shows the effect of these parameters on denitrification rate in the riparian zone. The increase of k means that denitrification occurs mostly in the soil layer near the surface. Therefore, if groundwater table is not high enough to interact with this soil layer, denitrification will rarely occur. On the other hand, a low k gives chances for denitrification to occur in the lower soil layers. Therefore, the lower k gives a higher average denitrification rate throughout the root zone. According to R_{max}, the increase of this parameter certainly raises the denitrification rate at every depth in the root zone. Root zone depth also affects denitrification process. A thicker root zone gives more chance for the groundwater to interact with the root zone, therefore, increases the nitrate removal by denitrification.

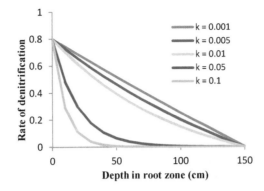

Figure 7.10 Effect of *k* on denitrification rate at different depth in the root zone

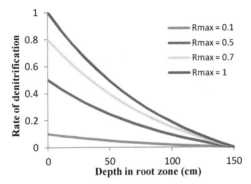

Figure 7.11 Effect of R_{max} on denitrification rate at different depth in the root zone

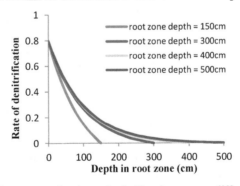

Figure 7.12 Effect of root zone depth on denitrification rate at different depth in the root zone

7.4.3 Testing with different scenarios

The modified SWAT was tested with four different scenarios which differ in the flow paths and nitrate sources from the upland area. The four scenarios are described as follows:

- *Scenario 1: Groundwater flow is the most significant flow to the riparian zone*

In this scenario, we decreased the value of *cn2* in order to reduce the generation of surface runoff, no tile drain was applied. Therefore, groundwater is the main contributor to the stream.

- *Scenario 2: Surface runoff is the dominating flow path*

In this scenario, we increased the value of *cn2* to increase the possibility of surface runoff generation, no tile drain was applied.

- *Scenario 3: Tile drains were applied in both upland and lowland areas which brought tile flow directly to the streams*

In this scenario, tile drains were applied in both upland and lowland areas. Tile drain were applied at the depth of 1 m, depth of impervious layer *dep_imp* was set at 3 m for the basin to allow perched water table to rise in order to generate tile flow.

- *Scenario 4: Tile drain was applied in the upland/ agricultural area but not in the lowland/riparian zone*

In this scenario, tile drain is only applied in the upland/agricultural area. *Dep_imp* was still set at 3 m. Flow and nitrate loads from upland area enter the lowland/riparian zone as additional input for hydrological and nitrogen processes.

Table 7.2 shows the results of flow and nitrate fluxes reaching the stream in 4 scenarios in the year 1999. The flow breakdown from different flow components including surface runoff, lateral flow, tile flow and groundwater flow are presented. Nitrate fluxes following various flow components are also calculated. In this test, we assume there is no denitrification in the upland area because tile drain is able to remove excess water and thus, groundwater table is low which does not create favorable condition for denitrification. Denitrification happens only in riparian zone which is estimated using the Riparian Nitrogen Model.

In general, it is clearly observed from table 7.2 that the total flow in scenario 2 and 3 was higher than the other two scenarios. Surface runoff and tile flow which are fast flow components contributed a high amount in scenario 2 and 3, respectively. Therefore, the total flow in scenarios 2 and 3 was higher than the other scenarios in which groundwater flow dominated. Nitrate fluxes differed in 4 scenarios following the difference in flow contribution. Nitrate flux in scenario 3 was the highest because tile flow brought nitrate from the soils directly to the river without any removal processes. On the other hand, nitrate flux in scenario 4 was the lowest because in this case, high tile flow from upland area entered lowland area and caused the rising of perched water table which created anaerobic condition and an interaction between nitrate in the soil and organic carbon in the rootzone, thus, denitrification occurred. The following discussion is specified for each scenario:

- In scenario 1, nitrate flux to the streams was mostly derived from groundwater flow. Denitrification process did not occur in the soil profile because of no interaction with the root zone. Nitrate was only lost by the removal processes in the shallow aquifer.

- In scenario 2, besides groundwater flow which was still the most significant contribution all the time of the year, surface runoff also contributed a high amount mostly in winter/high flow period when evapotranspiration is low. Compared to scenario 1, groundwater flow was lower; therefore, nitrate derived from groundwater flow also decreased. Although surface runoff accounts for a high percentage of the total flow, nitrate from this flow source was very limited because fertilizer was applied in the lower soil layers, not on the soil surface. Nitrate in this scenario was only lost from processes in the shallow aquifer.

- In scenario 3, tile flow was the dominating water component followed by groundwater flow. Therefore, nitrate load brought by tile flow accounted for the highest contribution. Still, there was very small nitrate removal from denitrification because the application of tile drains helped to remove exceeding water from the soil profile, thus, groundwater table was lower and had insignificant interaction with the root zone. Going through tile drains to the streams, nitrate fluxes had little interaction with the organic carbon in the soil; therefore, denitrification did not occur. Compared to two previous scenarios, the loss of nitrate in shallow aquifer was smaller because of the decrease of groundwater flow which resulted in the shrink of nitrate amount in shallow aquifer.

- In scenario 4, tile flow from upland agricultural fields was considered as an input for lowland hydrological processes. In the lowland area, because there was no tile drain to remove exceeding water, the high amount of tile flow input resulted in the rising of perched water table creating favorable conditions for denitrification to occur in the root zone. Therefore, denitrification was much higher than other scenarios.

In comparison with the original SWAT2005 (table 7.3), the SWAT_LS had a very slight difference in all flow components. In SWAT_LS, surface runoff tended to slightly decrease because a part of runoff from upland areas was infiltrated in lowland areas while groundwater flow slightly decreased because the routing between upland to lowland areas caused higher lag time for groundwater to reach the stream (in scenario 1, 2 and 3). Nitrate fluxes in the SWAT_LS model was lower than the original SWAT2005 because the lack of landscape routing between upland and lowland in the original SWAT2005 did not allow to simulate nitrate retention occurring during the routing process.

The difference in flow and nitrogen simulation between the two models was clearest in scenario 4. It is reminded that in this scenario, the upland area is considered as the agricultural area which is drained by tiles while lowland area is regarded as riparian zone. We assume there is no denitrification in the upland area because tile drain removes excess water and thus, groundwater table is low which does not create favorable condition for denitrification. In the original SWAT2005, the total flow result was much higher than the SWAT_LS model because flow retention in the riparian zone was not modeled with the lack of landscape process. Denitrification in this model remained 0 because no fertilizer was applied in riparian zone which did not give any input for nitrate removal process. However, denitrification occurred in the SWAT_LS model because riparian zone received a high amount of tile flow from the upland area which caused the rising of perched water table and created anaerobic condition for denitrification to happen.

Table 7.2 Water components and nitrate fluxes in different scenarios for testing the integrated wetland-SWAT_LS in year 1999

Components	Scenario1		Scenario 2		Scenario 3		Scenario 4	
	Flow (mm)	Nitrate flux (ton)	Flow (mm)	Nitrate flux (ton)	Flow (mm)	Nitrate flux (ton)	Flow (mm)	Nitrate flux (ton)
Surface runoff	19.76	0.04	129.87	0.42	30.50	0.06	53.56	0.08
Lateral flow	0.00	0.00	0.00	0.00	0.00	0.00	0.02	0.00
Tile flow	0.00	0.00	0.00	0.00	209.35	11.69	0.00 (From upland: 177.12)	0.00
Groundwater	338.51	14.70	252.86	13.71	131.75	5.55	262.01	7.65
Total	**358.27**	**14.74**	**382.73**	**14.13**	**371.60**	**17.30**	**315.59**	**7.73**
Loss of nitrate								
- By denitrification in the soil profile		0.00		0.00		0.005		8.34
- By denitrification in bank storage		0.00		0.00		0.00		0.00
- By processes in shallow aquifer		1.27		1.24		0.58		0.66

Table 7.3 Water components and nitrate fluxes in different scenarios by using original SWAT2005 in year 1999

Components	Scenario1		Scenario 2		Scenario 3		Scenario 4	
	Flow (mm)	Nitrate flux (ton)	Flow (mm)	Nitrate flux (ton)	Flow (mm)	Nitrate flux (kg/ha)	Flow (mm)	Nitrate flux (kg/ha)
Surface runoff	23.12	0.05	148.08	0.48	35.33	0.07	38.70	0.07
Lateral flow	0.08	0.001	0.07	0.001	0.15	0.002	0.15	0.002
Tile flow	0.00	0.00	0.00	0.00	205.59	11.69	177.12	11.67
Groundwater	345.24	15.82	249.04	14.74	135.22	5.79	161.98	5.80
Total flow	**368.44**	**15.87**	**397.19**	**15.22**	**376.29**	**17.55**	**377.95**	**17.54**
Loss of nitrate								
- By denitrification in the soil profile		0.00		0.00		0.00		0.00
- By denitrification in bank storage		Not calculated		Not calculated		Not calculated		Not calculated
- By processes in shallow aquifer		0.72		0.73		0.38		0.38

The details of the difference of nitrate balance in the two models can be seen in figure 7.13 and 7.14. Upland and lowland HRUs were separated in the SWAT2005 while flows from upland HRU were the inputs to lowland HRU in the SWAT_LS. Nitrate balances in the upland areas were similar in the two models while differences could be observed in the lowland area. With the additional input from the upland, there was higher amount of NO₃ in all processes happening in the lowland such as: higher NO₃ uptaken by plants, higher NO₃ percolated to the shallow aquifer, higher NO₃ produced from mineralization. As mentioned above, our concern is the nitrate removal by denitrification in the riparian zone which could be simulated in SWAT_LS when the riparian zone received a high amount of tile flow from the upland that resulted in high water table and anaerobic condition for denitrification to occur. While SWAT2005 was not able to simulate the nitrate removal by denitrification in the riparian zone, SWAT_LS predicted 8.3 ton NO₃ to be removed by denitrification in this case. Compared with the nitrate input from the soil profile in upland areas which was around 11.7 ton N mostly coming from tile flow, riparian zone was able to remove about 70% of the upland nitrate input. Therefore, denitrification in the riparian zone is a very important process that is necessary to be considered in the nitrate flux estimation. In addition to denitrification occurring in the soil profile, the loss of nitrate in the shallow aquifer was also higher in SWAT_LS because nitrate was removed in the upland shallow aquifer and then in lowland shallow aquifer as well.

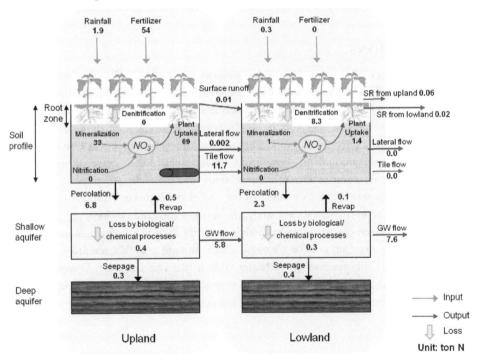

Figure 7.13 Nitrate balance in the hypothetical case study in scenario 4 using SWAT_LS

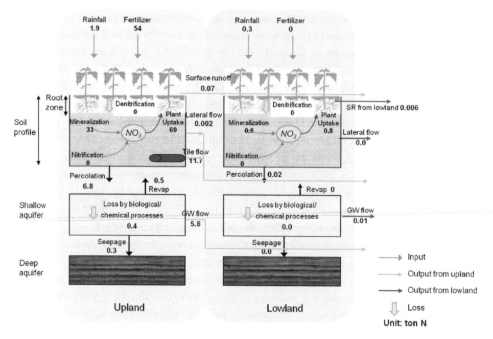

Figure 7.14 Nitrate balance in the hypothetical case study in scenario 4 using SWAT2005

7.5 CONCLUSIONS

Adding landscape variability and the routing process between upland to lowland landscape units in the SWAT model gives a better representation of hydrological and water quality processes in a river basin by setting up a relationship for flow and pollution fluxes between different landscape units in the river basin. In terms of nitrate simulation, the landscape approach does make a difference in modelling the denitrification process in case the lowland area receives a high amount of flow from upland which can cause a high perched groundwater table resulting in anaerobic conditions and interaction between groundwater with the organic matters in the root zone. In the test of the Riparian Nitrogen Model added in SWAT_LS for the hypothetical case study, the riparian zone did not have any effect when groundwater or surface runoff dominated while denitrification in the riparian zone occurred when it received a high amount of tile flow which brought a high amount of nitrate and causes the rising of the perched groundwater table in the riparian zone. Compared to the original SWAT2005 model, SWAT_LS is able to evaluate the efficiency of the riparian zone in nitrate removal by denitrification at the river basin scale.

Chapter 8

APPLICATION OF THE SWAT_LS MODEL IN THE ODENSE RIVER BASIN

8.1   MODEL SET UP FOR ODENSE RIVER BASIN USING SWAT_LS MODEL

Chapter 4 described the procedure of SWAT model setup for Odense river basin in detail. This chapter shows the application of the SWAT_LS model for the same case study. Only the differences between the two model setups are described as below.

HRU definition

The main difference in the two model setups is the addition of a landscape map in the HRU definition step. In the SWAT_LS model setup, HRU is defined by the overlay of land use, soil, slope maps and an additional landscape map.

With the objective of evaluating the effect of riparian zone on nitrate removal, the landscape map of the Odense river basin divides the basin into two landscape units: (i) upland areas which are dominated by agricultural areas and (ii) lowland areas which are the buffer zones along the river systems or riparian zones. This map was built based on the map of the distribution of organic soils of the basin (figure 8.1) assuming that organic soils located near the streams represent riparian zones. Assuming that the shortest width of the riparian zone is 50m on each side, the riparian zone map was created by overlaying the organic soil distribution map (figure 8.1) with a 50m buffer area along the river systems. Figure 8.2 illustrates the landscape map created for the Odense river basin.

In the user interface ARCSWAT, only land use and soil maps are overlaid in order to create HRUs. To take into account landscape variability in HRU generation, the landscape map was overlaid with the land use map to create a new land use map in which a new land use component called RIPR corresponding to riparian zones was added (table 8.1). The crop in RIPR land use is assumed to be grass only. The RIPR land use replaces a part of other land use areas, which changes the statistical percentage of crop distribution in the river basin resulting in changes in fertilizer applications. Therefore, in order to keep the same statistical figures for land use and crops, a modification was implemented in the new land use map in which some areas of grassland (GRAS) were randomly replaced by other crops. The total areas of grassland (GRAS) decreased was compensated by the area of riparian zones (RIPR) where grass is also grown. Table 8.1 presents the percentages of land use types and their crop rotations in the new land use map.

Figure 8.1 The map of organic soil distribution in the Odense river basin

Figure 8.2 The landscape map of the Odense river basin

Table 8.1 New types of land use and their crop rotations in Odense river basin

No.	Type of land use	Percentage of the basin	Crop rotation
1	Cattle farms	11.3	Spring Barley (year 1), Grass (year 2), Winter wheat (year 3), Maize (year 4)
2	Plant production	26.0	Spring Barley (year 1), Grass (year 2), Winter wheat (year 3 + year 4)
3	Pig farms	20.2	Spring Barley (year 1), Grass (year 2), Winter wheat (year 3), Winter barley (year 4)
4	Grass	14.6	Grass (year 1-4)
5	Coniferous forest	1.8	
6	Deciduous forest	8.1	
7	Urban area	7.5	50% impervious and 50% grass
8	Riparian zone	10.5	Grass

Tile drainage

For the upland areas which are dominated by agricultural areas, tile drain were applied in every HRU in agricultural areas which have the land use of cattle farms, plant production, pig farms or grass (table 8.1) at the depth of 1 m below the ground surface.

For the riparian zones, there are two cases:

- If tile drains are applied in the riparian zones, tile flow from the upland areas will enter tile drains in the riparian zones and go directly to the streams. In this case, drainage is the dominating source of flow.

- If tile drains are not applied in the riparian zones, tile flow from upland areas will be an input for the hydrological processes in the soil profile of riparian zones and contribute to the river through the groundwater flow path. In this case, stream flow is dominated by groundwater flow.

Therefore, the decision whether to apply tile drainage or not in a particular HRU in riparian zones depends on the dominating flow in the area that the HRU is located in. However, this information is rarely available for the whole river basin. From the field work of Banke (2005) at seven transects along two main streams of the Odense river basin (figure 8.3) and the study of Dahl et al. (2007) on evaluating the flow path distributions and identifying the dominant flow path through the riparian area to the stream, tile drainage was found to be the dominant flow at five transects (T1, T2, T4, T8, T9) while the other two transects (T6, T7) were dominated by groundwater flow (table 8.2 and figure 8.3). Based on this study's results, we assumed no tile drain is applied in the riparian zones approximately located between transect T8 to transect T4 (where transect T6 and T7 are located in between) along the main Odense river which corresponds to the riparian zones in sub-basin 17, 19 and 20. It is noted that the information about dominant riparian flow path is only available the river reach from T1 to T9 with seven studied transects. We assumed that this river reach and these 7 transects can represent the whole river basin which means 5 out of 7 (or 70%) riparian zones are dominated by tile drainage. Therefore, we randomly applied tile drains in the riparian zones so that the total percentage of riparian zones where tile drain is applied is approximately 70% of the total area of riparian zones in the whole river basin.

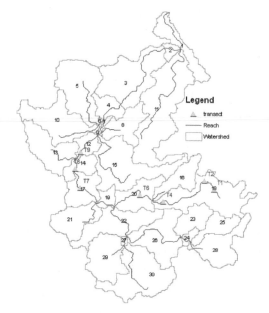

Figure 8.3 Locations of seven transects in the field work of Banke (2005)

Table 8.2 Riparian flow path types of transects in Odense river basin (Dahl et al., 2007)

Transect	Q_1	Q_2	Q_3	Q_4	Riparian flow path type
T1	Some	Some	-	Dominant	Drainage
T2	Some	Some	-	Dominant	Drainage
T4	Some	-	-	Dominant	Drainage
T6	Dominant	Some	-	-	Diffuse
T7	Some	-	Dominant	-	Direct
T8	Some	Some	-	Dominant	Drainage
T9	Some	-	-	Dominant	Drainage

Note: Q_1, Q_2, Q_3, Q_4 are referred to figure 8.4

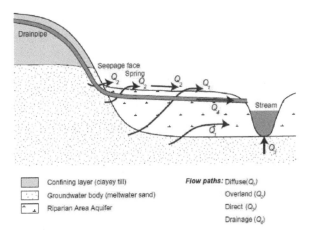

Figure 8.4 Flow paths through a riparian area to a stream

Fertilizer application

In the riparian zones, it was assumed that no fertilizer was applied. For other crops, fertilizer application was kept the same value as the SWAT setup built in chapter 4.

Denitrification in the riparian zone

As described in chapter 7, the Riparian Nitrogen Model was integrated in the SWAT_LS model in order to estimate the nitrate removal by denitrification in riparian zones. Denitrification rate is assumed to decrease with depth following equation 7.1. R_{max}, which is the maximum nitrate decay rate at the soil surface, and k, describing the rate at which the nitrate decay rate R declines with depth, are two parameters to be defined in this process.

A study from GEUS (Geological Survey of Denmark and Greenland) on a soil core taken from restored wetland called Brynemade located in Odense river basin has shown a decrease in denitrification rate through depth (unpublished data). From the results of nitrate removal at different depth taken every hour in the studied soil core, an equation representing the relationship between the nitrate reduction rate k and the depth below ground surface d was developed with the correlation coefficient at 0.77 (equation 8.1). However, this relationship only applies for the area below 10 cm from the ground surface. In the area from 0 - 10cm below the ground surface, the nitrate reduction rate is much higher and does not follow this relationship.

$$\log(k) = 0.0085d - 1.6751 \tag{8.1}$$

where k is nitrate reduction rate (1/hour), d is depth below the ground surface (negative value)

Assuming that this relationship can represent all riparian zones in Odense river basin, we calibrated R_{max} and k for the Riparian Nitrogen model and applied them for all riparian zones in the basin. The depth of rootzone r was set at 300 cm, same value with the studied soil core. The calibrated values for R_{max} and k are respectively 0.5 (1/day) and 0.02 (1/cm). It is clearly seen from figure 8.5 with the calibrated R_{max} and k, the RNM gave a very close estimation to denitrification rate of the soil core study. This calibrated RNM was applied to the soil layer of riparian zones lying below 10 cm from the ground surface. For the soil layer from 0 - 10 cm below surface, nitrate reduction rate was set at 0.8 day^{-1} based on the research of Revsbech et al. (2005) who studied the nitrogen transformation in a Danish flooded meadow and found that nitrate decreased exponentially with the rate of 0.8 day^{-1}.

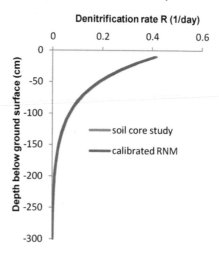

Figure 8.5 Denitrification rate in the calibrated Riparian Nitrogen model versus soil core study

8.2 COMPARISON BETWEEN THE SWAT_LS MODEL AND SWAT2005 MODEL IN FLOW AND NITROGEN SIMULATION

8.2.1 Comparison on flow simulation

The best parameter set that resulted from the calibration of the original SWAT set-up as described in chapter 4 was applied to the SWAT-LS set up for the Odense river basin in order to compare the performances of the two models. Figure 8.6 shows the predicted streamflow results of SWAT2005 and SWAT_LS versus measured flow. It is clearly observed that SWAT_LS gave lower values than SWAT2005 and also underestimated measurements. SWAT_LS decreased peak flows in high flow periods because two reasons: (i) a part of surface runoff from upland areas infiltrated back into riparian zones (ii) in some areas of riparian zones, tile drains were not applied which resulted in retention of tile flow from upland areas. In the description of SWAT_LS setup for Odense river basin, we assumed that 30% of riparian zones in the Odense river basin are not tile drained and dominated by groundwater flow, therefore, tile flow from upland areas has high retention when entering these riparian zones. Figure 8.7 also shows that the tile flow from the whole river basin simulated by SWAT_LS was lower than SWAT2005. In low flow periods, SWAT_LS also gave lower values because SWAT_LS gave a higher lag time for groundwater, which is the dominant flow path in these periods.

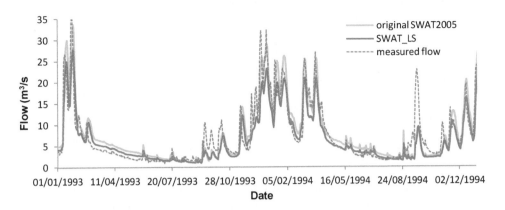

Figure 8.6 Comparison of flows simulated by SWAT_LS and original SWAT2005 versus measured flow at the gauging station 45_26 in the validation period

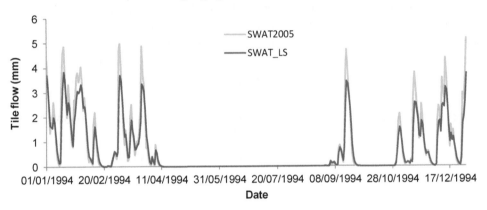

Figure 8.7 Comparison of tile flows simulated from SWAT2005 and SWAT_LS in the year 1994

In terms of NSE values at three gauging stations 45_26 (calibrated station in SWAT2005 model), 45_21 and 45_01, we can see that SWAT_LS also gave very good NSE at all three stations according to the model evaluation guidelines of Moriasi et al. (2007). Although using the calibrated parameter set from SWAT2005 setup, SWAT_LS gave a slightly better flow performance at the station 45_26 in both daily and monthly time steps with NSE at 0.84 and 0.91, respectively (table 8.3). The performance of SWAT_LS is also better at station 45_21, but worse at station 45_01 compared to SWAT2005.

Table 8.3 Comparison of Nash-Sutcliffe coefficients at gauging stations between SWAT_LS and original SWAT2005 in the period 1993-1998

Station	SWAT_LS		SWAT2005	
	Daily	Monthly	Daily	Monthly
45_26	0.84	0.91	0.82	0.90
45_21	0.83	0.90	0.75	0.88
45_01	0.74	0.75	0.83	0.90

Looking at the annual water balance between SWAT_LS and SWAT2005 in the period 1993-1998 in table 8.4, there are slight differences in distribution of flow components between the two models. Surface runoff was very slightly lower in SWAT_LS than SWAT2005, because surface runoff was not a dominating component in the Odense river basin. Tile flow was lower in SWAT_LS because of the retention of upland tile flow in some riparian zones in which tile drains were not applied. Groundwater flow was also lower in the SWAT_LS setup because the routing between landscape units gave a higher retention time for groundwater. Therefore, more water is kept in the river basin in SWAT_LS than in SWAT2005. Evapotranspiration only had a slight difference because of the difference in soil moisture that is available for evapotranspiration.

Table 8.4 Comparison of annual water balance between SWAT_LS and original SWAT2005 (1993-1998)

Water balance components	SWAT_LS (mm)	SWAT2005 (mm)
Precipitation	855	852
Surface runoff	51	52
Lateral soil	2	9
Tile flow	137	156
Groundwater flow	131	160
Evapotranspiration	486	461

8.2.2 Comparison on flow predictions with uncertainty between the two models

The objective of this section is to compare the SWAT_LS and SWAT2005 in terms of flow simulation taking into account parameter uncertainty. It is to evaluate how the modification in SWAT_LS affects the uncertainty of flow results. In this thesis, we used a simple pragmatic approach, based on Generalized Likelihood Uncertainty Estimation (GLUE) (Beven and Binley, 1992) to estimate uncertainty of discharge simulation. It is noted that GLUE is criticised by some researchers because it requires some subjective decisions (Hunter et al., 2005). On the other hand, GLUE is still widely used to estimate uncertainty in hydrological models (e.g. Montanari (2005), Winsemius et al. (2009), Krueger et al. (2010)). (Montanari, 2005). In this study, we are only concerned about parameter uncertainty.

The same parameters related to flow simulation that were used for calibration in chapter 4, were taken into account in the uncertainty analysis (see table 4.3 in chapter 4). Within the same ranges shown in table 4.3, the values of parameters were randomly distributed using Monte Carlo sampling techniques to generate 5000 parameter sets assuming that all parameters are uniformly distributed. Then, 5000 simulations corresponding to 5000 generated parameter sets were run with SWAT_LS and SWAT2005. Again, NSE was used as performance criterion and was calculated in each simulation. A threshold criterion of NSE at 0.5 was set for daily flow results to filter the 5000 simulations into behavioural and non-behavioural simulations (following the GLUE method). Behavioural models have NSE values greater or equals to 0.5 and the rest are non-behavioural models. Subsequently, the subset of behavioural models was used to estimate the uncertainty of discharge simulation while the subset of non-behavioural models was rejected. Within the GLUE framework (Beven and Binley, 1992), each behavioural model, i, is associated to a likelihood weight, L_i, ranging from 0 to 1, which is expressed as a function of the measure of fit, ε_i, of the behavioural models:

$$L_i = \frac{\varepsilon_i - \varepsilon_{min}}{\varepsilon_{max} - \varepsilon_{min}} \tag{8.2}$$

where ε_{max} and ε_{min} is the maximum and minimum measure of fit, ε_i is the measure of fit of behaviour model i. In this case, the measure of fit is NSE.

Then the likelihood weights are rescaled to a cumulative sum of 1 using the following equation:

$$w_i = \frac{L_i}{\sum_{i=1}^{n} L_i} \tag{8.3}$$

where w_i is the rescaled likelihood weight of behaviour model i, n is the number of behavioural models.

Then the weighted 5th, 50th and 95th percentiles, representing uncertainty bounds were computed for both SWAT_LS and SWAT2005 models. The estimates of the 5th and 95th percentiles of the cumulative likelihood distribution are chosen as uncertainty limits of the predictions for the two models.

The uncertainty analysis was implemented for both daily and monthly flow for the two models. Figure 8.8 shows the comparison between uncertainty bound of daily flow predictions in SWAT_LS and SWAT2005 while figure 8.9 illustrates the 50th percentile predicted flow of the two models versus observations. Figure 8.10 and figure 8.11 show the corresponding results for monthly predicted flow. Table 8.5 presents the values of several performance criteria to compare the two models.

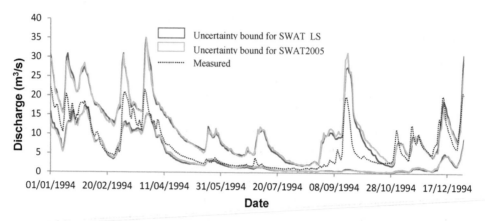

Figure 8.8 Comparison of uncertainty bounds for daily flow between SWAT_LS and SWAT2005

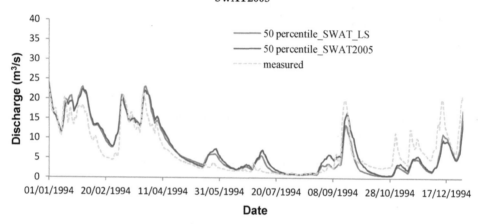

Figure 8.9 50th percentile predicted flow of SWAT_LS and SWAT2005 versus measured flow in daily time step

According to daily flow, we set the threshold of NSE at 0.5 to divide 5000 simulations into two subsets: behavioural and non-behavioural models. It is noted that the two models were run with the same Monte-Carlo parameter sets. From 5000 Monte-Carlo simulations, the number of SWAT_LS behavioural models satisfying NSE greater than 0.5 is 463 while it is only 168 in the SWAT2005 model (table 8.5). It implies that SWAT_LS performed better than SWAT2005 by giving higher probability to get a satisfactory representation of the modelled river basin.

The visual comparison between uncertainty bounds of SWAT_LS and SWAT2005 in the year 1994 in figure 8.8 shows that there was no big difference in uncertainty bounds although the number of behavioural models taken into account for this analysis was very different between the two models. The percentages of observations lying inside the uncertainty bounds were comparable between the two models (84-85%, table 8.5). These high percentages imply that both SWAT_LS and SWAT2005 are able to capture the important hydrological processes in the river basin. However, there were still a number of observations that fall outside the uncertainty bound implying that there may be some processes that were not taken into account or not well represented by the parameter ranges. Looking at the performance criteria in table 8.5, it is clearly observed that the values of NSE and correlation coefficients of 5th, 50th and 95th percentile flow were comparable between the two models. The low NSE of 95th percentile flow and the big difference between 95 percentile flow and observations especially in some peak flows imply that the two SWAT models could significantly overestimate the discharge while they were able to capture well the flow trend shown by the high correlation coefficients (table 8.5). The comparison of 50th percentile predicted discharges from the two models versus measured data show that the SWAT models were able to catch the hydrologic dynamic behaviour of the simulated river basin; however, they still did not capture perfectly the magnitudes of discharges at each time step.

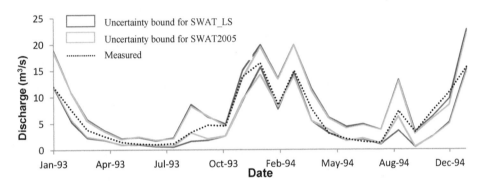

Figure 8.10 Comparison of uncertainty bounds for monthly flow between SWAT_LS and SWAT2005

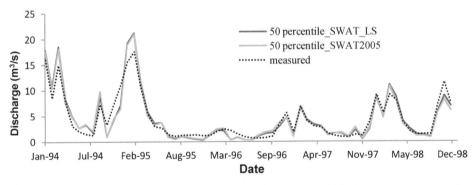

Figure 8.11 50th percentile predicted flow of SWAT_LS and SWAT2005 versus measured flow in monthly time step

In addition to daily flow, we also looked at uncertainty bounds of predicted monthly flows of the two models (figure 8.10). In case of monthly flow, we set the threshold of NSE at 0.75 to reject non-behavioural models in 5000 Monte-Carlo simulations. The threshold of NSE was chosen based on the model evaluation guidelines from (Moriasi et al., 2007) stating that NSE greater than 0.75 for monthly flow is a criterion for a "very good" model performance.

Similar to the case of daily flow, it is resulted that the number of behavioural models that are considered "very good" models predicted by SWAT_LS was much higher than SWAT2005 (316 versus 167). Moreover, within its uncertainty bound, SWAT_LS captured 85% of observations while SWAT2005 captured slightly smaller percentage at 76%. Therefore, it can be concluded that SWAT_LS performed better in flow simulation than SWAT2005 in both daily and monthly time steps.

Table 8.5 Compare uncertainties of SWAT_LS and SWAT2005 based on performance criteria

Performance criteria	Percentile	Daily		Monthly	
		SWAT_LS	SWAT2005	SWAT_LS	SWAT2005
Number of behavioral models Total simulation = 5000		463	168	316	167
NSE	max NSE	0.73	0.73	0.93	0.92
	min NSE	0.5	0.5	0.75	0.75
	5th percentile	0.46	0.46	0.79	0.81
	50th percentile	0.66	0.66	0.89	0.88
	95th percentile	-0.12	-0.12	0.65	0.70
Correlation coefficient	5th percentile	0.81	0.81	0.95	0.95
	50th percentile	0.86	0.85	0.96	0.95
	95th percentile	0.87	0.86	0.96	0.96
Average discharge Average measured discharge = 4.47 m³/s	5th percentile	2.25	2.26	3.12	3.22
	50th percentile	4.30	4.55	4.62	4.55
	95th percentile	7.37	7.38	6.25	6.01
Percentage of measurements lying in the uncertainty bound		84%	85%	85%	76%

The visualisation from figure 8.10 and the figures of different performance criteria at different percentiles in table 8.5 also show that there was no significant difference in uncertainty bounds between the two models. Compared to the daily flow, the 50th percentile predicted monthly flows in the two models seemed to not only catch the dynamic behaviours but also capture the magnitude quite well (figure 8.11). This is also reflected in the much better NSE of the predicted flow in all three percentiles for the monthly time step compared to the daily time step.

8.2.3 Comparison on nitrogen simulation

Similar to flow simulation, to make sure that the two models are compatible, the same values of nitrate related parameters resulted in the calibration of the original SWAT set-up in chapter 4 was applied in the SWAT_LS set up. To evaluate the effect of the Riparian Nitrogen Model added into SWAT_LS on nitrate fluxes, we assumed that denitrification only occurs in the riparian zones and is simulated by the Riparian Nitrogen Model. The upland areas are dominated by agricultural areas are drained by tiles to remove excess water and decrease groundwater table, thus, there is little interaction between groundwater with organic soils in the soil profile and denitrification rarely occurs. With this assumption, in the original SWAT2005 setup, there was no nitrate removal by denitrification while in SWAT_LS the denitrification process in riparian zones was taken into account by the Riparian Nitrogen Model.

Figure 8.12 shows the comparison of SWAT_LS and SWAT2005 versus measured nitrate fluxes in 2 year 1994 - 1995. It is clearly seen that both models were able to catch the variations of nitrate fluxes very well. However, both models seemed to overestimate the fluxes at peak points and underestimate at recession points. Compared to SWAT2005, SWAT_LS gave a lower result because of the contribution of denitrification in riparian zones in some sub-basins that were set to not connect with the stream through tile drains. Moreover, SWAT_LS allows higher lag time to groundwater flow. Therefore, there was more time for processes to remove nitrate occurring in the shallow aquifer.

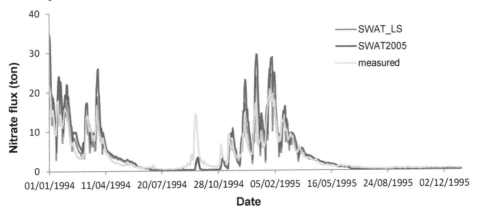

Figure 8.12 Comparison of nitrate fluxes simulated by SWAT_LS and original SWAT2005 versus measurements at the gauging station 45_21

Regarding the Nash-Sutcliffe efficiencies for nitrate fluxes at different gauging stations, the results show that SWAT_LS gave much higher values of NSE for both daily and monthly time steps compared to SWAT2005 (table 8.6). According to the guideline of performance evaluation from Moriasi et al. (2007), the results from SWAT_LS can be considered as "satisfactory" for daily nitrate fluxes and "good" for monthly results. Without denitrification, the SWAT2005 setup could not obtain a satisfactory fit to observations while SWAT_LS gave a much better fit with the inclusion of denitrification in riparian zones. Therefore, it is concluded that denitrification happening in riparian zones is a very important process in the Odense river basin. The contribution of denitrification and nitrate retention due to the flow retention that is considered in the landscape routing in SWAT_LS help to decrease the problem of overestimation of nitrate fluxes in the SWAT2005. In general it can be stated that the results become better with SWAT_LS.

Table 8.6 Comparison of Nash-Sutcliffe efficiencies for nitrate fluxes at gauging stations between SWAT_LS and original SWAT2005

Station	SWAT_LS		SWAT2005	
	Daily	Monthly	Daily	Monthly
45_21	0.55	0.71	0.15	0.35
45_26	0.55	0.69	0.31	0.43

Looking into details of the predicted nitrate fluxes and nitrate removal from different components between SWAT_LS and SWAT2005 in table 8.7, we can see that there is a big difference in the two models. Nitrate fluxes from surface runoff were compatible because of the similarity in surface runoff in the two models (see table 8.4 for flow results). The lateral flows were different between the two models because of the difference in lateral flow results. Lateral flow nitrate from the upland areas was used as input to riparian zones and was possibly removed in riparian zones, which is another reason of this distinction. The biggest difference is the nitrate fluxes from tile flow, which was around 250 ton/year. It is also seen in table 8.4 that tile flow in SWAT_LS was lower than SWAT2005 because of flow retention in riparian zones. With the retention of upland tile flow in riparian zones, groundwater table rose up creating favourable condition for denitrification. We also can observe in table 8.7 that denitrification in riparian zones was predicted at also around 250 ton/year which is compatible with the difference in tile flow nitrate fluxes between the two models. There was also a slight difference in groundwater nitrate fluxes because SWAT_LS increased the lag time of groundwater to reach the river by adding the landscape routing process. The higher lag time in groundwater also gave more time for nitrate removal processes in shallow aquifer. thus nitrate removal in shallow aquifer was higher in SWAT_LS.

Although denitrification in riparian zones helps to increase nitrate removal in the river basin, the contribution of riparian zones is not very significant as reported from a large number of studies on this field. From the total nitrate fluxes to the streams, which is around 2200 ton/year, denitrification only helps to remove more than 10%. However, it is reminded that more than 70% of riparian zones in the Odense river basin are drained by tile and dominated by tile drainage while only around 30% have retention capacity.

Table 8.7 Comparison of nitrate fluxes and nitrate removal between SWAT_LS and
SWAT2005

Nitrate components (unit: ton/year)	SWAT_LS	SWAT2005
Nitrate fluxes to the streams	*1894*	*2265*
- Through surface runoff	17	16
- Through lateral flow	1	22
- Through tile flow	1302	1536
- Through groundwater flow	574	691
Nitrate removal	*948*	*581*
- By denitrification in riparian zones	248	0
- By processes in shallow aquifer	700	581

8.3 EVALUATION OF DENITRIFICATION SIMULATED BY THE RIPARIAN NITROGEN MODEL

8.3.1 Sensitivity analysis for parameters of the Riparian Nitrogen Model

As described in chapter 7, the denitrification process estimated by the Riparian Nitrogen model is affected by three parameters shown in table 8.8. The effect of each parameter to the rate of denitrification was illustrated in section 7.4.2 of Chapter 7. In this section, we want to show the sensitivity of these parameters in which the effect of each parameter to the amount of nitrate removal by denitrification process is shown taking into account the simultaneous change of the remaining parameters. 500 parameter sets for the three parameters were randomly generated using Monte Carlo sampling approach. The ranges of the three parameters for Monte Carlo sampling are shown in table 8.8.

Table 8.8 Ranges of parameters of the Riparian Nitrogen Model for Monte Carlo sampling

No	Parameters	Definition	Unit	Range
1	R_{max}	Maximum denitrification rate at the soil surface	1/day	0.4-0.8
2	k_{denit}	Rate at which the denitrification rate declines with depth	1/mm	0.001-0.01
3	r	Depth of root zone	mm	1500-4000

In total 500 parameter sets were run with the SWAT_LS model of the Odense river basin. Previous results show that the riparian zones do not have significant effect on nitrate removal because a high percentage of the riparian area are connected to tile drain system which does not provide suitable condition for denitrification. Therefore, the nitrate removal capacity of riparian zones is not very significant in the present condition. In this section, we assume that the whole area of riparian zones is not connected to tile drains, thus, all tile flow from upland areas will not go directly to the river but have a retention impact from riparian zones. The model was run using meteorological data for 6 years from 1990-1995 in which the first three years were considered as warming-up period.

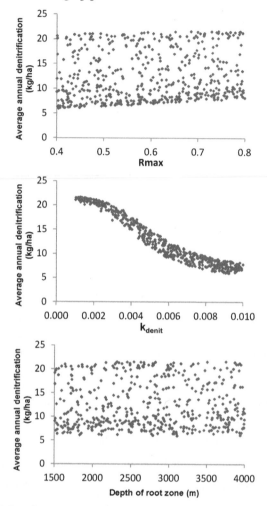

Figure 8.13 Sensitivity of parameters of the Riparian Nitrogen Model on average annual amount of nitrate removal by denitrification

Figure 8.13 shows the impact of changes in the parameters of the Riparian Nitrogen Model on the amount of nitrate removal by denitrification. This figure is plotted from the results of 500 Monte Carlo simulations; therefore, the impact of change in each parameter is also taken into account the simultaneous variations of the other two parameters. From the three dotty plots, it is clearly seen that in the assumption that the whole area of riparian zones is not drained, denitrification always happens and the amount of nitrate removal ranges from 6 to around 21.6 kg/ha for the whole basin or 367 to 1322 ton/year for the period 1993-1995. In addition, k_{denit} is the most sensitive parameters among the three parameters. With the variations of the other parameters, k_{denit} still shows a clear trend of the effect on the amount of nitrate removal while the other two parameters can result in very similar results with different values. It also can be said that k_{denit} shows a high level of identifiability while the other two parameters show low identifiabiltiy in terms of the amount of nitrate removal by denitrification.

8.3.2 Estimation of nitrate removal from denitrification with uncertainty

In the SWAT_LS model for Odense river basin, all denitrification parameters derived from the study of two soil cores are assumed to represent for all the riparian zones of the whole river basin. It is certainly known that the characteristic of riparian zones is not the same for the whole river basin and can differ dramatically in different places with distinct soil, crops and water levels. It is really difficult to identify all these characteristics for every riparian zone because of the complexity of water quality processes in such a complicated system like riparian zone. A model, even if it has a sub-module to simulate riparian zones, only is a simplistic representation for basic processes happening inside such a complicated system. The Riparian Nitrogen Model only focuses on the denitrification process which is the most important process in riparian zones. Considering at a catchment scale, it cannot give a certain result because of very limited information for the characteristics of riparian zones in the whole area. Therefore, in this thesis, we could not quantify with a high confidence the effect of riparian zones. However, assuming that the nitrate loads from upland areas is accurate, we can estimate the range of denitrification taking into account the uncertainty of denitrification parameters. Moreover, we also can observe the change of the amount of nitrate removal by denitrification when riparian zones in the whole area are not tile drained and help to trap and reduce flow and nitrogen.

The 500 Monte Carlo parameter sets, which resulted from the random distribution of 3 denitrification-related parameters between ranges shown in table 8.8, were run with SWAT_LS for two scenarios: (i) the present condition when around 70% of riparian zones are dominated by tile drainage, and (ii) the hypothetical condition when all riparian zones are not tile drained and have retention capacity for flow and nitrogen. The SWAT_LS model was run for two scenarios using the meteorological data from 1990-1995 in which the first three years were considered as warming-up period. Table 8.9 presents the amount of nitrate removal by denitrification considering parameter uncertainty in two scenarios. Figure 8.14 shows the difference in flow simulation between the two scenarios while figure 8.15 illustrates the change of denitrification uncertainty in monthly time step in two scenarios.

Table 8.9 Amount of denitrification considering uncertainty in different conditions

Scenarios	NO$_3$ input from upland areas		Amount of annual denitrification (1993-1995)	
	ton/year	kg/ha (based on the whole area of the river basin)	ton/year	kg/ha (based on the whole area of the river basin)
(i) Present condition			67 - 257	1.1 - 4.2
(ii) None of riparian zones are drained	1573	25.7	367 - 1322	6.0 - 21.6

Figure 8.14 shows that when all riparian zones are not tile drained, peak flows decreased due to the flow retention of riparian zones while there was an increase in magnitude and dynamic variations of flow in the low flow period due to the increase of surface runoff from the saturation of riparian zones. Figure 8.15 shows that the uncertainty bound of scenario (ii) when all riparian zones in the whole basin are not tile drained was broader than the uncertainty bound of the present condition because denitrification happened in a larger area. Looking at the relationship between flow and denitrification (figure 8.14 and 8.15), it is seen that the denitrification only happened in the high flow periods and there was no denitrification in the low flow periods. In the low flow periods, the water level was too low to reach the soil profile, thus, no interaction between base flow and organic matter in the soil happened. In the high flow periods, riparian zones received a high amount of tile flow from upland areas which increased the water level and resulted in an interaction between base flow and organic carbon causing denitrification. From figure 8.15, we can observe that there was a time delay between the occurrence of denitrification occurring and the input of nitrate from upland areas. At the beginning, flow brought nitrate to the riparian zones; however, the amount of flow was not enough for the water to rise up to organic soil layers. Then, denitrification occurred where there is high water level that caused interaction between nitrate and organic matters. However, when input from upland areas decreased, there was still denitrification occurring because the retention of flow in riparian zones was able to keep the water level high while the inflows are decreasing.

From table 8.9 and figure 8.15, we can see that at the present condition, denitrification does not have a significant impact on nitrate removal taken into account parameter uncertainty because only a small area of riparian zones has flow and nitrate retention and removal capacity. The nitrate removal is only about 4~17%. However, when none of the riparian zones are drained, they can really perform their retention function by which the effectiveness of riparian zones for nitrate removal increases dramatically resulting in a nitrate removal of around 25~85% taking into account parameter uncertainty.

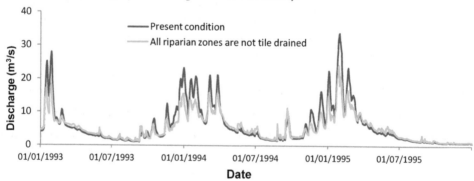

Figure 8.14 Comparison on flow simulation between two scenarios of riparian zones

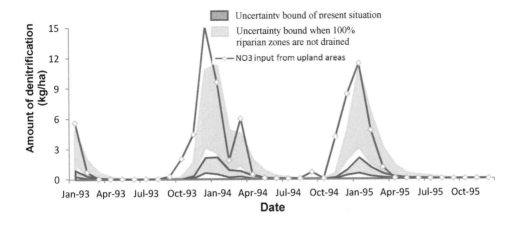

Figure 8.15 Uncertainty on amount of monthly denitrification between two scenarios of riparian zones

8.3 CONCLUSION

The application of all SWAT modifications in the Odense river basin showed that compared to the original SWAT2005, SWAT_LS did give improvements in the simulation of flow and nitrate fluxes based on evaluating the Nash-Sutcliffe coefficients. The comparison between the two models taking into account parameter uncertainty by running 5000 Monte-Carlo simulations showed that SWAT_LS had a much higher number of parameter sets that gave satisfactory performances (behavioural models) in both daily and monthly time steps. It implies that SWAT_LS performed better than SWAT2005 by giving higher probability to get a satisfactory representation of the modelled river basin. Although the big difference in the number of behavioural models, uncertainty bounds are compatible between the two models. It was also shown that in the Odense river basin, riparian zones do not have a significant effect on nitrate removal because a large area of riparian zones in the area are drained and dominated by tile drainage.

Because of the limited measurements on denitrification and the assumption that all riparian zones in the Odense river basin have the same characteristics, the estimation of denitrification took into account the denitrification-related parameter uncertainty in (i) the current situation and (ii) a hypothetical condition when all riparian zones are not drained. The results showed that the uncertainty bound of scenario (ii) is broader than the uncertainty bound of the present condition because denitrification happens in a larger area. In the present condition, the nitrate removal is only about 4~17% taking into account uncertainty. However, if all riparian zones in the area are not drained and can really perform their retention function, the effectiveness of the riparian zones on nitrate removal will increase dramatically to around 25~85%.

Chapter 9

CONCLUSIONS AND RECOMMENDATIONS

9.1 CONCLUSIONS

9.1.1 Summary of objectives and methodology

The riparian zone, which is the interface between terrestrial and aquatic ecosystems, plays an important role in nitrogen removal in river basins. Despite the minor proportion of the land area that it covers, its major role in nutrient removal has been verified in a large number of studies related to the effect of wetlands/riparian zones in river basins and streamflow catchments. Nitrogen removal is mostly achieved by denitrification which favours anaerobic conditions created by high water levels. Most of the studies related to the effect of riparian zones on nitrogen removal are limited to small scales and fieldwork. Very limited research has been carried on modelling their effects at larger scales or at river basin/catchment scales.

Nowadays, there are many river basin models available that are able to provide predictions of pollutant loading from diffuse sources. Several models are available in simulating hydrologic and chemical processes in wetlands/riparian zones. However, there are limited studies on integrating wetland/riparian zone models with river basin models in order to evaluate the effect of wetland/riparian zones at river basin scales. The SWAT model is a well known and broadly used model that can simulate hydrological and nutrient transport processes, as shown by the large number of SWAT application reviewed by Gassman et al. (2007). In terms of riparian zone modelling, SWAT contains a sub-module VFS for estimating flow and pollutant retention in buffer strips based on empirical equations derived from observations.

The main objective of this thesis is to evaluate the effect of riparian zones on nitrate removal at the river basin scale using the SWAT model. This thesis focuses on modifying the SWAT model by (i) adding routing features across different landscape units and (ii) adding the Riparian Nitrogen Model (RNM) to simulate denitrification processes in riparian zones. The first modification aims at taking into account the landscape position of each Hydrological Response Unit (HRU) and creating relationships between HRUs in upland and HRUs in lowland landscape units, which is considered important for realistically modelling flow and pollutant transport processes. Following from the first modification, the second modification introduces a new model for simulating denitrification processes in HRUs that contain riparian zones. In this model, the denitrification rate is assumed to decline with depth and the denitrification process is activated below the groundwater table. This interaction occurs via two mechanisms: (i) groundwater passing through the riparian buffer before discharging into the stream, and (ii) surface water being temporarily stored within the riparian soils during flood events.

By implementing these modifications in the SWAT modelling suite, this thesis intends to contribute to further developments and improvements of SWAT. The main case study in this thesis is the Odense river basin in Denmark, which is an agricultural-dominated area and a densely tile-drained catchment. This case study was chosen by the EU AQUAREHAB project because of its comprehensive source of data being available, and the interest in the effects of wetland restoration in the river basin scale which had been heavily modified for agricultural purposes in the recent past.

The main work carried out in this thesis can be summarised as follows:

- A SWAT model was built for the Odense river basin to simulate flow and nitrogen transport processes; an evaluation of its performance in this specific case was carried out.

- The SWAT model was compared with an existing DAISY- MIKE SHE model in terms of flow and nitrogen simulations. These two models have different model structures and use different concepts for flow and nitrogen. Moreover, different SWAT models with different structures were developed and compared with each other. An evaluation of the performance of the different models and model structures was implemented in this work.

- An approach to take into account landscape variability was introduced in SWAT. This corresponds to the first modification of the SWAT model as mentioned above, called SWAT_LS. In this version, the landscape position is taken into account when creating HRUs for the SWAT model. In addition, flow and nitrogen can be routed from upland to lowland HRUs before reaching the streams. This approach considers the effect of hydrological and nitrogen processes in upland areas on processes in lowland areas and takes into account the interaction between upland and lowland areas. A comparison between SWAT_LS and the original SWAT model was carried out to assess the change in flow and nitrogen results caused by this modification. SWAT_LS was tested in a very simple hypothetical case study to explore water balance and transport processes in the modified approach.

- The Riparian Nitrogen Model (RNM) was added to SWAT_LS, which is the second modification mentioned above. RNM requires the groundwater table in each HRU as a necessary input to define the area where denitrification occurs in the riparian zone. A procedure to predict the groundwater table from SWAT soil moisture results as introduced by Vazquez-Amábile and Engel (2005) was added to SWAT by modifying the code. This modification was also tested with a simple hypothetical case study to evaluate the effect of the Riparian Nitrogen Model in different scenarios.

- Finally, the modified SWAT model SWAT_LS was applied to the Odense river basin. The performance of this model was evaluated by comparing with measured data and the original SWAT model. The riparian zone was determined based on the distribution map of organic soils. It was assumed that the area of the organic soil nearby the streams and buffer areas of 50 m from the streams belong to riparian zone. The effect of riparian zones in the Odense river basin was evaluated from the modelling results. The uncertainty on flow and nitrate removal by denitrification was also investigated for this case study.

9.1.2 Summary of main conclusions and contributions

✓ *Performance of SWAT model on flow and nitrogen simulations*

The SWAT model performed well in replicating the daily streamflow hydrograph at the calibrated station in the validation periods, although some of the peak flows were either under- or over-predicted and variations in low flows were not captured well. SWAT also gave good results at the two stations downstream and upstream of the calibrated station. The performance of SWAT was better for the monthly streamflow predictions as compared to the daily predictions, which is consistent with previous reviews of many SWAT studies. SWAT predicted that tile drainage and groundwater are both dominant flow components. Surface runoff gave a small contribution to streamflow and lateral flow was insignificant. The SWAT result is compatible with the field work of Banke (2005) and findings of Dahl et al. (2007) stating that the dominant flow path is tile flow at five studied transects and groundwater flow at two others, and that surface runoff is not the dominant flow path.

In terms of nitrogen simulation, SWAT was able to replicate the correct trend and magnitude of nitrate fluxes versus observations. However, it did not accurately capture many of the daily fluxes well, especially several of the peak nitrate fluxes during the high flow periods (from November to April). Similar to the flow results, the monthly results for nitrate fluxes predicted by SWAT were closer to the observations and can be considered 'satisfactory' for monthly time step results, according to the model evaluation guidelines of Moriasi et al. (2007). Breaking down the nitrate fluxes into different flow components, it was shown that tile flow was the most significant source of nitrate followed by groundwater flow. Surface runoff did not bring significant nitrate loads to the river because fertilizers are assumed to be applied in the lower soil layer beneath the ground surface. In general, the performance of SWAT on nitrogen simulation was not as good as on flow simulation, based on Nash-Sutcliffe as the performance indicator. However, pollutant transport and transformation processes are usually very complicated and cannot be represented perfectly in a simplified model, which is why water quality modelling usually does not obtain good results as flow simulation. The SWAT predicted results compared well by similarity in magnitude and variation of predicted versus measured daily nitrogen fluxes, which implies that SWAT can be an effective tool for simulating nitrogen loadings in the Odense River basin.

✓ *Comparison between SWAT and DAISY-MIKE SHE models*

The results of the SWAT model were compared with the DAISY-MIKE SHE model in terms of discharge and nitrogen simulation. Results from the SWAT model were compared with the DAISY-MIKE SHE model set up from Van der Keur et al. (2008) taking into account the uncertainty of soil hydraulic properties and slurry parameters. The results of 24 runs of the DAISY-MIKE SHE model provided an uncertainty bound for the flow and nitrate fluxes at the gauging station used for comparison. For the flow simulation, the SWAT results fitted quite well within the uncertainty range of the DAISY-MIKE SHE values. Almost all the SWAT values were within range in the high flow period while some values were smaller than the corresponding DAISY-MIKE SHE minimum values in the low flow period. Compared to the 50th percentile (median) flow ranked DAISY-MIKE SHE outputs from the 24 simulations, the SWAT model gave a better fit to the measured values than the DAISY-MIKE SHE model as indicated by the higher values for Nash-Sutcliffe efficiency and correlation coefficient. When comparing the annual average water balance between the two models, a striking water balance difference was shown on flow component breakdown. Although both models predicted that subsurface flow is dominant, the proportion between tile flow and groundwater flow are very different between the two models. Virtually all of the subsurface flow occurred as tile flow in DAISY-MIKE SHE while subsurface flow inputs were roughly evenly split between tile flow and groundwater flow in SWAT. While SWAT predicted tile flow to occur only in the high flow period and groundwater is the main source to streamflow in low flow periods, DAISY-MIKE SHE still predicted tile flow to be generated as the main source for low flow periods. It is noted that there are no direct field observations or measurements of tile flow in the low flow periods in the Odense River basin. However, it is clear that the two models responded very differently during the low flow periods, and it is likely that the DAISY-MIKE SHE provided a more accurate representation because of the physics-based concepts used in the DAISY-MIKE SHE model.

In terms of nitrogen simulation, the SWAT-predicted nitrate flux also fitted well within the uncertainty bound of the 24 DAISY-MIKE SHE model simulations, especially in the high flow periods, which implies that SWAT simulates river basin nitrogen processes reasonably well. Comparing the annual average nitrate fluxes along different pathways between the two models, there a big difference was found in the predicted tile flow and groundwater flow nitrate fluxes due to the difference in flow predictions. Although tile flow and groundwater flow are both dominant flow paths, nitrate fluxes from tile flow are the dominant source. The reason is that tile flow is a fast component that can bring nitrate loads directly to the river without any removal processes, while groundwater is a slow component which allows time for nitrate removal processes occurring in the shallow aquifer, as simulated in SWAT.

✓ *Performance of different model structures in modelling flow and nitrogen fluxes*

A comparison was carried out between different SWAT setups including (i) SWAT without tile drainage, (ii) SWAT with tile drainage. It was shown that when using the same parameter set for the two models which were derived from a rough manual calibration of model (i), the flow result for setup (ii) was significantly better than for setup (i). The inclusion of tile drain parameters improved the SWAT model performance especially in high flow periods by producing higher peak flows and steeper hydrograph recession while there was not significant change during low flow periods that are dominated by groundwater flow. It can be concluded that tile drainage has an important contribution to streamflow and is a very important component to be included in the model of the Odense river basin.

The two models were calibrated using the auto-calibration tool. A surprising result was obtained in that the SWAT setup without tile drainage simulation was found to achieve a much better fit than the SWAT setup with tile drainage. The SWAT model without tile drainage generated a high amount of surface runoff that could easily catch the peak flow in high flow periods and also small variations in low flow periods of measured data. However, it is known that most of the case study area is drained by tiles and tile flow is the most important source that provides water to the streams and very little surface runoff is observed. Based on this information, the SWAT model without tile drainage in which surface runoff is the primary flow component is considered unrealistic since it is neglecting the importance of tile drainage in the system.

From the comparison between the two different SWAT model setups, and the comparison between SWAT and DAISY-MIKE SHE, it is concluded that different model structures and different model concepts can obtain similar results that fit reasonably well to the observations but give very different results for flow paths and pollutant fluxes. These results show the equifinality problem of over-parameterized models in which many different parameter sets will give almost identical fits to the measured data, but can yield dramatically different predictions of how the system will behave as conditions change. It is clear that statistical metrics, which are important for evaluating the goodness-of-fit of simulated results to measured data, are nevertheless unable to fully distinguish whether a model is capable of determining water and pollutant pathways correctly. Therefore, it is extremely important to select a suitable model structure for a particular case study based on all possible information about the area, expert knowledge, observations in the field, etc.

✓ *The effect of adding the approach to represent the landscape variability in the SWAT model*

The approach to represent the landscape variability in SWAT (SWAT_LS) gave two modifications to the SWAT model: (i) dividing a sub-basin into two landscape units: upland and lowland, and (ii) allowing hydrological routing between them. The results showed that the sensitivities of flow-related parameters on flow response in SWAT_LS were similar to their behaviours in the original SWAT model. The flow value for each flow component in SWAT_LS had a small change compared to the original SWAT. Consequently, it can be concluded that the added routing between landscape units can affect the water volume at the outflow of the catchment, but does not influence the flow behaviour within the sub-basins. Curve number *cn2* and depth of impervious layer *dep_imp* are the most sensitive parameters not only to flow response in each model but also to flow differences between the two models, i.e. SWAT_LS and SWAT2005. In comparison with SWAT2005, SWAT_LS decreases surface runoff because a part of surface runoff from the upland areas is infiltrated back into the lowland areas, decreasing groundwater flow because of longer lag time for this component moving across two landscape units before reaching the river. Tile flow predicted in SWAT_LS has a very small increase compared to SWAT2005 because the amount of water infiltrated back into the lowland areas results in a slightly higher amount of tile flow generated in these areas. Moreover, the proportion between upland and lowland areas seems to have a very strong effect on upland flows; however, the effect on flows from the whole sub-basin is not significant. With this approach, it is possible to take into account the landscape position for each HRU, and therefore, it is possible to evaluate flow and nitrogen results for each landscape unit. Moreover, this approach also represents the interaction between upslope HRUs and downslope HRUs, which gives a better representation for the hydrological processes in reality.

✓ *The effect of the Riparian Nitrogen Model in modelling denitrification in riparian zones*

The Riparian Nitrogen Model brought a new concept in modelling the denitrification process in riparian zones in the SWAT model. With the added landscape approach (SWAT_LS), it is possible to differentiate HRUs that belong to riparian zones, and therefore it becomes possible to apply the Riparian Nitrogen model only in the riparian zone HRUs and evaluate the results. It is reminded that in the Riparian Nitrogen Model, the denitrification process is simulated via two mechanisms: (i) groundwater passing through the riparian buffer before discharging to the stream and (ii) surface water being temporarily stored within the riparian soils during flood event. The mechanism (ii) likely has a less contribution because of the low frequency of occurrence compared to mechanism (i). In the test of the integrated SWAT_Riparian zone model in the hypothetical case study, the riparian zone did not have any effect when deep groundwater or surface runoff dominated while denitrification in the riparian zone occurred when the riparian zone received high amounts of tile flow which brought a high amount of nitrate and caused the rising of the perched groundwater table in the riparian zone. Compared to the original SWAT2005 model, SWAT_LS is able to evaluate the efficiency of the riparian zone in nitrate removal by denitrification at the river basin scale.

✓ *Application of SWAT_LS in simulating hydrological processes and evaluating the nitrate removal in riparian zones by denitrification in the Odense river basin*

The application of all SWAT modifications in the Odense river basin showed that compared to the original SWAT2005, SWAT_LS did give improvements in the simulation of flow and nitrate fluxes as indicated by the Nash-Sutcliffe coefficients. A comparison between the two models taking into account parameter uncertainty by running 5000 Monte-Carlo simulations showed that SWAT_LS had a much higher number of parameter sets that gave satisfactory performances (behavioural models) at both daily and monthly time steps. This implies that SWAT_LS performed better than SWAT2005 by having a higher probability of getting a satisfactory representation of the modelled river basin. Despite the big difference in the number of behavioural models, the uncertainty bounds are comparable between the two models. It was also shown that in the Odense river basin, riparian zones do not have a significant effect on nitrate removal because a large number of riparian zones in the area are drained and dominated by tile drainage.

Because of the limited measurements on denitrification and the assumption that all riparian zones in the Odense river basin have the same characteristics, the estimation of denitrification took into account the denitrification-related parameter uncertainty in (i) the current situation and (ii) a hypothetical condition when all riparian zones are not drained. The results showed that the uncertainty bound of scenario (ii) is broader than the uncertainty bound of the present condition because denitrification happens in a larger area. In the present condition, the nitrate removal is only about 4~17% taking into account uncertainty. However, if all riparian zones in the area are not drained and can really perform their retention function, the effectiveness of the riparian zones on nitrate removal will increase dramatically to around 25~85%.

9.2   RECOMMENDATIONS

The modifications in the SWAT_LS model introduced in this study are expected to give a contribution to a better representation of the hydrological and nitrogen removal processes, particularly denitrification, by including landscape routing processes and introducing the Riparian Nitrogen Model to the SWAT model.

The applied case study for the Odense river basin which is an agricultural river basin covered by a densely tile drainage networks, provided a suitable area for testing all the modifications developed in this thesis. However, further research should be done for other case studies with different characteristics to test the effect of landscape routing processes in different conditions. The Odense river basin is a very flat area dominated by sub-surface flow, therefore the landscape routing processes do not have a significant effect on flow response although it helps to give a better representation of the hydrological processes in the river basin. However, it is possible that landscape routing may have a considerable effect in other river basins that are dominated by surface runoff. This should be verified by further research.

The addition of landscape routing in SWAT_LS, although it gives a better description of the hydrological processes, also increases the number of parameters in the model. In this thesis, a river basin is divided in two landscape units: upland and lowland, and therefore the number of parameters related to flow routing is increased. Here, we do not mean parameters that relate to soil or land use or any catchment characteristics, we mean the user-input parameters that are related to flow routing.

Some examples of these parameters include *surlag* which is the parameter related to surface runoff lag time, *gw_delay* which represents the delay of groundwater from soil profile to shallow aquifer, *alpha_bf* which is the baseflow recession constant, *gdrain* related to the lag time of tile flow. In this thesis, we assume that these parameters take the same value for both upland areas and lowland areas to decrease the over-parameterised problem. However, because upland and lowland areas have very different areas and distinct characteristics, these parameters should have different values for different landscape units. To prevent adding more parameters to the model, relationships between these parameters in upland and lowland areas should be developed.

In this thesis, a river basin is only divided into two landscape units. However, the landscape approach can be applied to more landscape units for a river basin that has significant change in topography or landscape characteristics. This can also be considered in the further applications of the landscape approach added to SWAT. As mentioned above, a relationship between parameters in different landscape units should be developed to prevent over-parameterisation.

The Riparian Nitrogen Model added a new concept to simulate the denitrification process in the SWAT model and was applied successfully in the Odense river basin. However, it is really difficult to verify the result of denitrification at the river basin scale. There were very few measurements and they were limited in plot scale. Consequently, to verify the results, more studies and measurements should be carried out. Moreover, in this thesis it was assumed that the characteristics of riparian zones such as soil, land use, denitrification rate, etc. are the same for the whole river basin. There was also limited information about the dominant flow in riparian zones. This lack of information resulted in uncertainty on denitrification estimations. Therefore, more information and more field research on the Odense basin case study and on other river basins will be useful for further evaluation.

REFERENCES

Aber, J.D., C.L. Goodale, S.V. Ollinger, M.-L. Smith, A.H. Magill, M.E. Martin, R.A. Hallett, and J.L. Stoddard. 2003. Is Nitrogen Deposition Altering the Nitrogen Status of Northeastern Forests? Bioscience 53(4):375-389.

Abrahamsen, P., and S. Hansen. 2000. Daisy: an open soil-crop-atmosphere system model. Environmental Modelling & Software 15(3):313-330.

Arheimer, B., and H.B. Wittgren. 2002. Modelling nitrogen removal in potential wetlands at the catchment scale. Ecological Engineering 19(1):63-80.

Arnold, J.G., P.M. Allen, M. Volk, J.R. Williams, and D.D. Bosch. 2010. Assessment of different representations of spatial variability on SWAT model performance. Transaction of ASABE 53(5):1433-1443.

Arnold, J.G., and N. Fohrer. 2005. SWAT2000: current capabilities and research opportunities in applied watershed modelling. Hydrological Processes 19(3):563-572.

Arnold, J.G., R. Srinivasan, R.S. Muttiah, and J.R. Williams. 1998. Large area hydrologic modeling and assessment part 1: Model development. JAWRA Journal of the American Water Resources Association 34(1):73-89.

Band, L.E., D.L. Peterson, S.W. Running, J. Coughlan, R. Lammers, J. Dungan, and R. Nemani. 1991. Forest ecosystem processes at the watershed scale: basis for distributed simulation. Ecological Modelling 56(0):171-196.

Band, L.E., C.L. Tague, P. Groffman, and K. Belt. 2001. Forest ecosystem processes at the watershed scale: hydrological and ecological controls of nitrogen export. Hydrological Processes 15(10):2013-2028.

Banke, M. 2005. Method for estimating flow path distribution in stream valleys (in Danish). M.Sc. thesis. University of Copenhagen, Copenhagen, Denmark

Beasley, D.B., and L.F. Huggins. 1981. ANSWERS, areal nonpoint source watershed environment response simulation: user's manual. U.S. Environmental Protection Agency, Region V, Great Lakes National Program OfficeChicago, I11. (USA).

Bergström, L., and N. Brink. 1986. Effects of differentiated applications of fertilizer N on leaching losses and distribution of inorganic N in the soil. Plant Soil 93(3):333-345.

Beven, K., and A. Binley. 1992. The future of distributed models: Model calibration and uncertainty prediction. Hydrological Processes 6(3):279-298.

Bicknell, B.R., J.C. Imhoff, J.L. Kittle, T.H. Jobes, and A.S. Donigian. 2000. Hydrological Simulation Program–Fortran: HSPF, Version 12 User's Manual. Athens: US Environmental Protection Agency.

Birkinshaw, S.J., and J. Ewen. 2000. Nitrogen transformation component for SHETRAN catchment nitrate transport modelling. Journal of hydrology 230(1-2):1-17.

Borah, D.K., and M. Bera. 2003. Watershed-scale hydrologic and nonpoint-source pollution models: Review of mathematical bases. Transaction of ASAE 46(6):1553-1566.

Bouraoui, F., and B. Grizzetti. 2008. An integrated modelling framework to estimate the fate of nutrients: Application to the Loire (France) Ecological Modelling 212(3-4):450-459.

Boussinesq, J. 1872. Therorie des ondes et des remous qui se propagent le long d'un canal rectangulaire horizontal,. Journal de Mathématiques Pures et Appliquées 7:55-108.

Boyer, E., C. Goodale, N. Jaworski, and R. Howarth. 2002. Anthropogenic nitrogen sources and relationships to riverine nitrogen export in the northeastern U.S.A. In The Nitrogen Cycle at Regional to Global Scales, 137-169. E. Boyer, and R. Howarth, eds: Springer Netherlands.

Boyer, E.W., R.B. Alexander, W.J. Parton, C. Li, K. Butterbach-Bahl, S.D. Donner, R.W. Skaggs, and S.J.D. Grosso. 2006. Modeling denitrification in terrestrial and aquatics ecosystems at regional scales. Ecological Applications 16(6):2123-2142.

Brown, L.C., and J. T.O. Barnwell. 1987. The enhanced water quality models QUAL2E and QUAL2E-UNCAS documentation and user manual. Athens, GA: USEPA.

Buresh, R.J., M.E. Casselman, W.H. Patrick Jr, and N.C. Brady. 1980. Nitrogen Fixation in Flooded Soil Systems, A Review. In Advances in Agronomy, 149-192. Academic Press.

Burt, T.P., and B.P. Arkell. 1987. Temporal and spatial patterns of nitrate losses from an agricultural catchment. Soil Use and Management 3(4):138-142.

Campbell, D.H., J.S. Baron, K.A. Tonnessen, P.D. Brooks, and P.F. Schuster. 2000. Controls on nitrogen flux in alpine/subalpine watersheds of Colorado. Water Resources Research 36(1):37-47.

Chavan, P., and K. Dennett. 2008. Wetland Simulation Model for Nitrogen, Phosphorus, and Sediments Retention in Constructed Wetlands. Water, Air, and Soil Pollution 187(1-4):109-118.

Conan, C.l., F.a. Bouraoui, N. Turpin, G. de Marsily, and G. Bidoglio. 2003. Modeling Flow and Nitrate Fate at Catchment Scale in Brittany (France). Journal of Environmental Quality 32(6):2026-2032.

Dahl, M., B. Nilsson, J.H. Langhoff, and J.C. Refsgaard. 2007. Review of classification systems and new multi-scale typology of groundwater-surface water interaction. Journal of hydrology 344(1-2):1-16.

Daniel, E.B., J.V. Camp, E.J. LeBoeuf, J.R. Penrod, J.P. Dobbins, and M.D. Abkowitz. 2011. Watershed modeling and its applications: A state-of-the-art review. The open hydrology journal 5:26-50.

Daniels, R.B., and J.W. Gilliam. 1996. Sediment and Chemical Load Reduction by Grass and Riparian Filters. Soil Science Society of America Journal 60(1):246-251.

Del Grosso, S., D. Ojima, W. Parton, A. Mosier, G. Peterson, and D. Schimel. 2002. Simulated effects of dryland cropping intensification on soil organic matter and greenhouse gas exchanges using the DAYCENT ecosystem model. Environmental Pollution 116, Supplement 1(0):S75-S83.

Del Grosso, S.J., A.R. Mosier, W.J. Parton, and D.S. Ojima. 2005. DAYCENT model analysis of past and contemporary soil N₂O and net greenhouse gas flux for major crops in the USA. Soil and Tillage Research 83(1):9-24.

DeLaney, T.A. 1995. Benefits to downstream flood attenuation and water quality as a result of constructed wetlands in agricultural landscapes. Journal of Soil and Water Conservation 50(6):620-626.

Douglas-Mankin, K.R., R. Srinivansan, and J.G. Arnold. 2010. Soil and Water Assessment Tool (SWAT) model: Current developments and applications. Transactions of the ASABE 53(5):1423-1431.

Driscoll, C.T., G.B. Lawrence, A.J. Bulger, T.J. Butler, C.S. Cronan, C. Eagar, K.F. Lambert, G.E. Likens, J.L. Stoddard, and K.C. Weathers. 2001. Acidic Deposition in the Northeastern United States: Sources and Inputs, Ecosystem Effects, and Management Strategies. Bioscience 51(3):180-198.

Du, B., J.G. Arnold, A. Saleh, and D.B. Jaynes. 2005. Development and application of SWAT to landscapes with tiles and potholes. Transactions of the ASABE 48(3):1121-1133.

Du, B., A. Saleh, D.B. Jaynes, and J.G. Arnold. 2006. Evaluation of SWAT in simulating nitrate nitrogen and atrazine fates in a watershed with tiles and potholes. Transactions of the ASABE 49(4):949-959.

EC. 2000. Directive 2000/60/EC of the European Parliament and of the Council of 23 October 2000 establishing a framework for Community action in the field of water policy. http://eurlex.europa.eu/LexUriServ/LexUriServ.do?uri=OJ:L:2000:327:0001:0072:EN:PDF (accessed 1 Sep 2011).

El-Nasr, A.A., J.G. Arnold, J. Feyen, and J. Berlamont. 2005. Modelling the hydrology of a catchment using a distributed and a semi-distributed model. Hydrological Processes 19(3):573-587.

Environment Centre Odense. 2007. Odense Pilot River Basin. Pilot project for river basin management planning. Water Framework Directive Article 3. Danish Ministry of the Environment - Environment Centre Odense.

Ewen, J., Parkin, G., O'Connell, E.,. 2000. SHETRAN: distributed river basin flow and transport modeling system. Journal of Hydrologic Engineering 5(3):250–258.

Ferrant, S., F. Oehler, P. Durand, L. Ruiz, J. Salmon-Monviola, E. Justes, P. Dugast, A. Probst, J.-L. Probst, and J.-M. Sanchez-Perez. 2011. Understanding nitrogen transfer dynamics in a small agricultural catchment: Comparison of a distributed (TNT2) and a semi distributed (SWAT) modeling approaches. Journal of hydrology 406(1–2):1-15.

Fyns county. 2003. Odense Pilot River Basin. Provisional Article 5 Report pursuant to the Water Framework Directive. Fyns county. http://www.helpdeskwater.nl/publish/pages/13970/voorwoordensamenvatting.pdf (accessed 1 Sep 2011).

Galloway, J.N., F.J. Dentener, D.G. Capone, E.W. Boyer, R.W. Howarth, S.P. Seitzinger, G.P. Asner, C.C. Cleveland, P.A. Green, E.A. Holland, D.M. Karl, A.F. Michaels, J.H. Porter, A.R. Townsend, and C.J. Vöosmarty. 2004. Nitrogen Cycles: Past, Present, and Future. Biogeochemistry 70(2):153-226.

Gassman, P.W., J.G. Arnold, R. Srinivasan, and M. Reyes. 2010. The worldwide use of the SWAT model: Technological driver, networking impacts, and simulation trends. Transactions of the ASABE.

Gassman, P.W., M.R. Reyes, C.H. Green, and J.G. Arnold. 2007. The Soil and Water Assessment Tool: Historical development, applications, and future research directions. Transactions of the ASABE 50(4):1211-1250.

Geyer, D.J., C.K. Keller, J.L. Smith, and D.L. Johnstone. 1992. Subsurface fate of nitrate as a function of depth and landscape position in Missouri Flat Creek watershed, U.S.A. Journal of Contaminant Hydrology 11(1–2):127-147.

Gold, A.J., P.A. Jacinthe, P.M. Groffman, W.R. Wright, and R.H. Puffer. 1998. Patchiness in Groundwater Nitrate Removal in a Riparian Forest. Journal of Environmental Quality 27(1):146-155.

Green, C.H., M.D. Tomer, M. Di Luzio, and J.G. Arnold. 2006. Hydrologic evaluation of the soil and water assessment tool for a large tile-drained watershed in Iowa. Trans. ASABE 49(2):413-422.

Gregory, S.V., F.J. Swanson, and W.A. McKee. 1991. An ecosystem perspective on riparian zones Bioscience 41:540-551

Greve, M.H., and H. Breuning-Madsen. 2005. Soil Mapping in Denmark.: Office for Official Publications of the European Communities Luxembourg.

Groffman, P.M., M.A. Altabet, J.K. Böhlke, K. Butterbach-Bahl, M.B. David, M.K. Firestone, A.E. Giblin, T.M. Kana, L.P. Nielsen, and M.A. Voytek. 2006. Methods for measuring denitrification: diverse approaches to a difficult problem. Ecological Applications 16(6):2091-2122.

Groffman, P.M., E.A. Axelrod, J.L. Lemunyon, and W.M. Sullivan. 1991. Denitrification in Grass and Forest Vegetated Filter Strips. Journal of Environmental Quality 20(3):671-674.

Groffman, P.M., A.J. Gold, and R.C. Simmons. 1992. Nitrate Dynamics in Riparian Forests: Microbial Studies. Journal of Environmental Quality 21(4):666-671.

Groffman, P.M., and G.C. Hanson. 1997. Wetland Denitrification: Influence of Site Quality and Relationships with Wetland Delineation Protocols. Soil Science Society of America Journal 61(1):323-329.

Groffman, P.M., G. Howard, A.J. Gold, and W.M. Nelson. 1996. Microbial Nitrate Processing in Shallow Groundwater in a Riparian Forest. Journal of Environmental Quality 25(6):1309-1316.

Groffman, P.M., J.M. Tiedje, G.P. Robertson, and S. Christensen. 1988. Denitrification at different temporal and geographical scales: proximal and distal controls. Advances in nitrogen cycling in Agricultural Ecosystems:174-192.

Güntner, A., and A. Bronstert. 2004. Representation of landscape variability and lateral redistribution processes for large-scale hydrological modelling in semi-arid areas. Journal of hydrology 297(1–4):136-161.

Hansen, J.R., J.C. Refsgaard, V. Ernstsen, S. Hansen, M. Styczen, and R.N. Poulsen. 2009. An integrated and physically based nitrogen cycle catchment model. Hydrology Research 40.4:347-364.

Hansen, J.R., J.C. Refsgaard, S. Hansen, and V. Ernstsen. 2007. Problems with heterogeneity in physically based agricultural catchment models. Journal of hydrology 342(1-2):1-16.

Hansen, S. 1984. Estimation of Potential and Actual Evapotranspiration. Nordic Hydrology 15(4 - 5):205 - 212.

Hansen, S., H.E. Jensen, N.E. Nielsen, and H. Svendsen. 1991. Simulation of nitrogen dynamics and biomass production in winter wheat using the Danish simulation model DAISY. Nutrient Cycling in Agroecosystems 27(2):245-259.

Hattermann, F.F., V. Krysanova, A. Habeck, and A. Bronstert. 2006. Integrating wetlands and riparian zones in river basin modelling. Ecological Modelling 199(4):379-392.

Haycock, N.E., and T.P. Burt. 1993. The sensitivity of rivers to nitrate leaching: The effectiveness of near-stream land as a nutrient retention zone. In Landscape sensitivity, 261-272. D.S.G. Thomas, and R.J. Allison, eds. London: Wiley.

Henriksen, H.J., L. Troldborg, P. Nyegaard, T.O. Sonnenborg, J.C. Refsgaard, and B. Madsen. 2003. Methodology for construction, calibration and validation of a national hydrological model for Denmark. Journal of hydrology 280(1-4):52-71.

Hill, A.R. 1996. Nitrate Removal in Stream Riparian Zones. Journal of Environmental Quality 25(4):743-755.

Horrigan, L., R.S. Lawrence, and P. Walker. 2002. How sustainable agriculture can address the environmental and human health harms of industrial agriculture. Environmental Health Perspectives 110(5):445-456.

Howard-Williams, C. 1985. Cycling and retention of nitrogen and phosphorus in wetlands: a theoretical and applied perspective. Freshwater biology 15:391 - 431.

Howarth, R.W., G. Billen, D. Swaney, A. Townsend, N. Jaworski, K. Lajtha, J.A. Downing, R. Elmgren, N. Caraco, T. Jordan, F. Berendse, J. Freney, V. Kudeyarov, P. Murdoch, and Z. Zhao-Liang. 1996. Regional nitrogen budgets and riverine N & P fluxes for the drainages to the North Atlantic Ocean: Natural and human influences. In Nitrogen Cycling in the North Atlantic Ocean and its Watersheds, 75-139. R. Howarth, ed: Springer Netherlands.

Huang, Z., B. Xue, and Y. Pang. 2009. Simulation on stream flow and nutrient loadings in Gucheng Lake, Low Yangtze River Basin, based on SWAT model. Quaternary International 208(1-2):109-115.

Huber, W.C., and R.E. Dickinson. 1988. Storm-Water Management Model, Version 4. Part a: user's manual.

Hunter, N.M., P.D. Bates, M.S. Horritt, A.P.J. De Roo, and M.G.F. Werner. 2005. Utility of different data types for calibrating flood inundation models within a GLUE framework. Hydrology and Earth System Science 9(4):412-430.

Imhol, J.G., J. Fitzgibbon, and W.K. Annable. 1996. A hierarchical evaluation system for characterizing watershed ecosystems for fish habitat. Canadian Journal of Fisheries and Aquatic Sciences 53(S1):312-326.

Jordan, T.E., D.L. Correll, and D.E. Weller. 1993. Nutrient Interception by a Riparian Forest Receiving Inputs from Adjacent Cropland. Journal of Environmental Quality 22(3):467-473.

Kadlec, R.H., and R.L. Knight. 1996. Treatment wetlands. CRC Press, Boca Raton, FL.

Kadlec, R.H., and S.D. Wallace. 2008. Treatment wetlands.

Kang, M.S., S.W. Park, J.J. Lee, and K.H. Yoo. 2005. Applying SWAT for TMDL programs to a small watershed containing rice paddy fields. Agricultural Water Management 79(1):72-92.

Kazezyılmaz-Alhan, C.M., M.A. Medina, and C.J. Richardson. 2007. A wetland hydrology and water quality model incorporating surface water/groundwater interactions. Water Resources Research 43(4):W04434.

Keddy, P.A. 2000. Wetland Ecology - Principles and conservation. Cambridge University Press.

Kirchner, J.W. 2006. Getting the right answers for the right reasons: Linking measurements, analyses, and models to advance the science of hydrology. Water Resource Research 42(3):W03S04.

Knisel, W. 1980. CREAMS, a field scale model for chemicals, runoff, and erosion from agricultural management systems. USDA Conservation Research Report No. 26.

Knowles, R. 1982. Denitrification. Microbiological Reviews 46(1):28.

Krueger, T., J. Freer, J.N. Quinton, C.J.A. Macleod, G.S. Bilotta, R.E. Brazier, P. Butler, and P.M. Haygarth. 2010. Ensemble evaluation of hydrological model hypotheses. Water Resources Research 46(7):W07516.

Lam, Q.D., B. Schmalz, and N. Fohrer. 2011. The impact of agricultural Best Management Practices on water quality in a North German lowland catchment. Environmental Monitoring and Assessment 183(1-4):351-379.

Leonard, R.A., W.G. Knisel, and D.A. Still. 1987. GLEAMS: groundwater loading effects of agricultural management systems. Transactions of the ASABE 30(5):1403-1418.

Li, C., J. Aber, F. Stange, K. Butterbach-Bahl, and H. Papen. 2000. A process-oriented model of N_2O and NO emissions from forest soils: 1. Model development. Journal of Geophysical Research: Atmospheres 105(D4):4369-4384.

Li, C., S. Frolking, and T.A. Frolking. 1992. A model of nitrous oxide evolution from soil driven by rainfall events: 1. Model structure and sensitivity. Journal of Geophysical Research: Atmospheres 97(D9):9759-9776.

Li, C., V. Narayanan, and R.C. Harriss. 1996. Model estimates of nitrous oxide emissions from agricultural lands in the United States. Global Biogeochemical Cycles 10(2):297-306.

Line, D.E., D.L. Osmond, L.A. Lombardo, G.L. Grabow, D.E. Wise-Frederick, J. Spooner, and K. Hall. 2002. Nonpoint sources. Water Environ. Res. 74(5):1-41.

Martin, T., N.K. Kaushik, J.T. Trevors, and H.R. Whiteley. 1999. Review: Denitrification in temperate climate riparian zones. Water, Air, and Soil Pollution 111(1-4):171-186.

McClain, M.E., E.W. Boyer, C.L. Dent, S.E. Gergel, N.B. Grimm, P.M. Groffman, S.C. Hart, J.W. Harvey, C.A. Johnston, E. Mayorga, W.H. McDowell, and G. Pinay. 2003. Biogeochemical Hot Spots and Hot Moments at the Interface of Terrestrial and Aquatic Ecosystems. Ecosystems 6(4):301-312.

McDowell, W., W. Bowden, and C. Asbury. 1992. Riparian nitrogen dynamics in two geomorphologically distinct tropical rain forest watersheds: subsurface solute patterns. Biogeochemistry 18(2):53-75.

Meselhe, E.A., E.H. Habib, O.C. Oche, and S. Gautam. 2009. Sensitivity of conceptual and physically based hydrologic models to temporal and spatial rainfall sampling Journal of Hydraulic Engineering 14(7).

Middlebrooks, E.J., and A. Pano. 1983. Nitrogen removal in aerated lagoons. Water Research 17(10):1369-1378.

Mitsch, W.J., and J.G. Gosselink. 2000. Wetlands. Wiley, New York, NY.

Montanari, A. 2005. Large sample behaviors of the generalized likelihood uncertainty estimation (GLUE) in assessing the uncertainty of rainfall-runoff simulations. Water Resources Research 41(8):W08406.

Monteith, J.L. 1965. Evaporation and environment. *In* State and Movement of Water in Living Organisms: Proc. 19th Symp.Society of Experimental Biology, 205-234. Cambridge, U.K.: Cambridge University Press.

Moreno-Mateos, D., C. Pedrocchi, and F.A. Comín. 2010. Effects of wetland construction on water quality in a semi-arid catchment degraded by intensive agricultural use. Ecological Engineering 36(5):631-639.

Moriasi, D.N., J.G. Arnold, M.W. Van Liew, R.L. Bingner, R.D. Harmel, and T.L. Veith. 2007. Model evaluation guidelines for systematic quantification of accuracy in watershed simulations. Transactions of the ASABE 50(3):885-900.

Muñoz-Carpena, R., J.E. Parsons, and J.W. Gilliam. 1999. Modeling hydrology and sediment transport in vegetative filter strips. Journal of hydrology 214(1–4):111-129.

Nash, J.E., and J.V. Sutcliffe. 1970. River flow forecasting through conceptual models part I -- A discussion of principles. Journal of hydrology 10(3):282-290.

Nasr, A., M. Bruen, P. Jordan, R. Moles, G. Kiely, and P. Byrne. 2007. A comparison of SWAT, HSPF and SHETRAN/GOPC for modelling phosphorus export from three catchments in Ireland. Water Research 41(5):1065-1073.

Neitsch, S.L., J.G. Arnold, J.R. Kiniry, and J.R. Williams. 2004. Soil and Water Assessment tool Input/Output file documentation version 2005. Texas Water Resources Institute, College Station. Texas.

Neitsch, S.L., J.G. Arnold, J.R. Kiniry, J.R. Williams, and K.W. King. 2005. Soil and Water Assessment tool theoretical documentation version 2005. http://swatmodel.tamu.edu/media/1292/swat2005theory.pdf (accessed 1 June 2011). Texas Water Resources Institute, College Station. Texas.

Nielsen, K., M. Styczen, H.E. Andersen, K.I. Dahl-Madsen, J.C. Refsgaard, S.E. Pedersen, J.R. Hansen, S.E. Larsen, R.N. Poulsen, B. Kronvang, C.D. Børgesen, M. Stjernholm, K. Villholth, J. Krogsgaard, V. Ernstsen, O. Jørgensen, J. Windolf, A. Friis-Christensen, T. Uhrenholdt, M.H. Jensen, I.S. Hansen, and L. Wiggers. 2004. Odense Fjord—Scenarios for Reduction of Nutrients (in Danish). NERI Technical Report no. 485. Danish National Environmental Research Institute (NERI). Roskilde, Denmark.

Olivera, F., M. Valenzuela, R. Srinivasan, J. Choi, H. Cho, S. Koka, and A. Agrawal. 2006. ARCGIS-SWAT: A geodata model and GIS interface for SWAT. The Journal of the American Water Resources Association 42(2):295-309.

Osborne, L.L., and D.A. Kovacic. 1993. Riparian vegetated buffer strips in water-quality restoration and stream management. Freshwater biology 29(2):243-258.

Parkin, T.B. 1987. Soil Microsites as a Source of Denitrification Variability1. Soil Science Society of America Journal 51(5):1194-1199.

Parton, W.J., A.R. Mosier, D.S. Ojima, D.W. Valentine, D.S. Schimel, K. Weier, and A.E. Kulmala. 1996. Generalized model for N_2 and N_2O production from nitrification and denitrification. Global Biogeochemical Cycles 10(3):401-412.

Penman, H.L. 1956. Evaporation: An introductory survey. Netherlands Journal of Agricultural Science 4:7-29.

Pinay, G., L. Roques, and A. Fabre. 1993. Spatial and temporal patterns of denitrification in a riparian forest. Journal of Applied Ecology 30:581-591.

Pohlert, T., J.A. Huisman, L. Breuer, and H.G. Frede. 2005. Modelling of point and non-point source pollution of nitrate with SWAT in the river Dill, Germany. Advances in Geosciences 5:7-12.

Rabalais, N.N. 2002. Nitrogen in Aquatic Ecosystems. AMBIO: A Journal of the Human Environment 31(2):102-112.

Ramachandra, T.V., N. Ahalya, and C. Rajasekara Murthy. 2005. Aquatic ecosystems: conservation, restoration and management. Capital Publishers, New Delhi.

Rassam, D.W., D.E. Pagendam, and H.M. Hunter. 2005. The Riparian Nitrogen Model (RNM) - Basic theory and conceptualisation. CRC for Catchment Hydrology Technical Report.

Rassam, D.W., D.E. Pagendam, and H.M. Hunter. 2008. Conceptualisation and application of models for groundwater - surface water interactions and nitrate attenuation potential in riparian zones. Environmental Modelling & Software 23(7):859-875.

Reddy, K. 1982. Nitrogen cycling in a flooded-soil ecosystem planted to rice (Oryza sativa L.). Plant and Soil 67(1):209-220.

Reddy, K.R., W.H. Patrick, and F.E. Broadbent. 1984. Nitrogen transformations and loss in flooded soils and sediments. Critical Reviews in Environmental Control 13(4):273 - 309.

Refsgaard, J.C., and H.H.J. Henriksen. 2004. Modelling guidelines - terminology and guiding principles. Advances in Water Resources 27(1):71-82.

Refsgaard, J.C., and B. Storm. 1995. MIKE SHE. *In* Computer models of watershed Hydrology, 809 - 846. V.P. Singh, ed: Water Resources Publications, Highlands Ranch, CO.

Refsgaard, J.C., B. Storm, and T. Clausen. 2010. Système Hydrologique Europeén (SHE): review and perspectives after 30 years development in distributed physically-based hydrological modelling. Hydrology Research 41(5):355-377.

Revsbech, N.P., J.P. Jacobsen, and L.P. Nielsen. 2005. Nitrogen transformations in microenvironments of river beds and riparian zones. Ecological Engineering 24(5):447-455.

Roth, N., J.D. Allan, and D. Erickson. 1996. Landscape influences on stream biotic integrity assessed at multiple spatial scales. Landscape Ecology 11(3):141-156.

Rücker, K., and J. Schrautzer. 2010. Nutrient retention function of a stream wetland complex--A high-frequency monitoring approach. Ecological Engineering 36(5):612-622.

Salvetti, R., M. Acutis, A. Azzellino, M. Carpani, C. Giupponi, P. Parati, M. Vale, and R. Vismara. 2008. Modelling the point and non-point nitrogen loads to the Venice Lagoon (Italy): the application of water quality models to the Dese-Zero basin Desalination 226(1-3):81-88.

Sather, J.H., and R.D. Smith. 1984. An overview of major wetland functions and values. Western Energy and Land Use Team, U.S. Fish and Wildlife service, Washington, DC.

Savenije, H.H.G. 2010. Topography driven conceptual modelling (FLEX-TOPO). Hydrology and Earth System Sciences 14:2681-2692.

Schilling, K., and C. Wolter. 2009. Modeling Nitrate-Nitrogen Load Reduction Strategies for the Des Moines River, Iowa Using SWAT. Environ. Manage. 44(4):671-682.

Schnabel, R.R., J.A. Shaffer, W.L. Stout, and L.F. Cornish. 1997. Denitrification Distributions in Four Valley and Ridge Riparian Ecosystems. Environmental Management 21(2):283-290.

Schoumans, O.F., M. Silgram, D.J.J. Walvoort, P. Groenendijk, F. Bouraoui, H.E. Andersen, A. Lo Porto, H. Reisser, G. Le Gall, S. Anthony, B. Arheimer, H. Johnsson, Y. Panagopoulos, M. Mimikou, U. Zweynert, H. Behrendt, and A. Barr. 2009. Evaluation of the difference of eight model applications to assess diffuse annual nutrient losses from agricultural land. Journal of Environmental Monitoring 11(3):540-553.

Seitzinger, S., J.A. Harrison, J.K. Böhlke, A.F. Bouwman, R. Lowrance, B. Peterson, C. Tobias, and G.V. Drecht. 2006. Denitrification across landscapes and waterscapes: a synthesis. Ecological Applications 16(6):2064-2090.

Sharpley, G.W., and J.R. Williams. 1990. EPIC - erosion/productivity impact calculator: 1. model documentation. Technical Bulletin No. 1768. U.S. Department of Agriculture, Temple, Texas, USA.

Shepherd, B., D. Harper, and A. Millington. 1999. Modelling catchment-scale nutrient transport to watercourses in the U.K. Hydrobiologia 395-396(0):227-238.

Skaggs, R.W. 1980. DRAINMOD reference report. Fort Worth, Tex: U.-S.S.N.T. Center.

Skaggs, R.W. 1999. Drainage simulation models. *In* Agricultural drainage. R.W. Skaggs, and J. van Schilfgaarde, eds. madison, Wisconsin, USA: Soil Science Society of America.

Stisen, S., M.F. McCabe, J.C. Refsgaard, S. Lerer, and M.B. Butts. 2011. Analyzing parameter sensitivities using remotely 1 sensed surface temperature - towards spatial calibration of complex distributed hydrological models. Journal of Hydrology (accepted).

Styczen, M., and B. Storm. 1993a. Modelling of N-movements on catchment scale - a tool for analysis and decision making. 1. Model description. Fertilizer research 36(1):1-6.

Styczen, M., and B. Storm. 1993b. Modelling of N-movements on catchment scale - a tool for analysis and decision making. 2. A case study. Fertilizer research 36(1):7-17.

Sueker, J.K., D.W. Clow, J.N. Ryan, and R.D. Jarrett. 2001. Effect of basin physical characteristics on solute fluxes in nine alpine/subalpine basins, Colorado, USA. Hydrological Processes 15(14):2749-2769.

Sui, Y., and J.R. Frankengerger. 2008. Nitrate loss from subsurface drains in an agricultural watershed using SWAT2005. Transactions of the ASABE 51(4):1263-1272.

Thompson, J.R., H.R. Sørenson, H. Gavin, and A. Refsgaard. 2004. Application of the coupled MIKE SHE/MIKE 11 modelling system to a lowland wet grassland in southeast England. Journal of hydrology 293:151-179.

Townsend, A.R., R.W. Howarth, F.A. Bazzaz, M.S. Booth, C.C. Cleveland, S.K. Collinge, A.P. Dobson, P.R. Epstein, E.A. Holland, D.R. Keeney, M.A. Mallin, C.A. Rogers, P. Wayne, and A.H. Wolfe. 2003. Human health effects of a changing global nitrogen cycle. Frontiers in Ecology and the Environment 1(5):240-246.

Trepel, M., and L. Palmeri. 2002. Quantifying nitrogen retention in surface flow wetlands for environmental planning at the landscape-scale. Ecological Engineering 19(2):127-140.

Troldborg, L. 2004. The influence of conceptual geological models in the simulation of flow and transport in Quaternary aquifer systems, PhD thesis, Geological Survey of Denmark and Greeland, Report 2004/107, GEUS, Copenhagen. http://www.fiva.dk/doc/thesis/ltr_phd_thesis.pdf (accessed 9 April 2012)

Tuppad, P., K.R. Douglas-Mankin, T. Lee, R. Srinivansan, and J.G. Arnold. 2011. Soil and Water Assessment Tool (SWAT) hydrologic/water quality model: Extended capability and wider adoption. Transactions of the ASABE 54(5):1677-1684.

Tuppad, P., N. Kannan, R. Srinivasan, C. Rossi, and J. Arnold. 2010. Simulation of Agricultural Management Alternatives for Watershed Protection. Water Resources Management 24(12):3115-3144.

Ullrich, A., and M. Volk. 2009. Application of the Soil and Water Assessment Tool (SWAT) to predict the impact of alternative management practices on water quality and quantity. Agricultural Water Management 96(8):1207-1217.

USDA-NRCS. 2004. Chapter 10: Estimation of direct runoff from storm rainfall. *In* NRCS National Engineering Handbook, Part 630: Hydrology. Washington, D.C.: U.S. Department of Agriculture, Natural Resources Conservation Service. http://directives.sc.egov.usda.gov/viewerFS.aspx?hid=21422 (accessed 8 May 2012).

Vagstad, N., H.K. French, H.E. Andersen, H. Behrendt, B. Grizzetti, P. Groenendijk, A. Lo Porto, H. Reisser, C. Siderius, J. Stromquist, J. Hejzlar, and J. Deelstra. 2009. Comparative study of model prediction of diffuse nutrient losses in response to changes in agricultural practices. Journal of Environmental Monitoring 11(3):594-601.

van Breemen, N., E.W. Boyer, C.L. Goodale, N.A. Jaworski, K. Paustian, S.P. Seitzinger, K. Lajtha, B. Mayer, D. van Dam, R.W. Howarth, K.J. Nadelhoffer, M. Eve, and G. Billen. 2002. Where did all the nitrogen go? Fate of nitrogen inputs to large watersheds in the northeastern U.S.A. Biogeochemistry 57-58(1):267-293.

Van den Heuvel, R.N., M.M. Hefting, N.C.G. Tan, M.S.M. Jetten, and J.T.A. Verhoeven. 2009. N_2O emmision hotspots at different spatial scales and governing factors for small scale hotspots.

van der Keur, P., J.R. Hansen, S. Hansen, and J.C. Refsgaard. 2008. Uncertainty in Simulation of Nitrate Leaching at Field and Catchment Scale within the Odense River Basin. Vadose Zone Journal 7(1):10-21.

Van Griensven, A., and W. Bauwens. 2003. Concepts for river water quality processes for an integrated river basin modelling. Water Science and Technology 48(3):1-8.

Vazquez-Amábile, G.G., and B.A. Engel. 2005. Use of SWAT to compute groundwater table depth and streamflow in the Muscatatuck river watershed. Transactions of the ASAE 48(3):991-1003.

Volk, M., S. Liersch, and G. Schmidt. 2009. Towards the implementation of the European Water Framework Directive?: Lessons learned from water quality simulations in an agricultural watershed. Land Use Policy 26(3):580-588.

Vought, L.B.M., J. Dahl, C.L. Pedersen, and J.O. Lacoursiere. 1994. Nutrient retention in riparian ecotones. Ambio 23(6).

Vymazal, J. 1995. Algae and element cycling in wetlands. Lewis Publishers Inc.

Vymazal, J. 2001. Type of constructed wetlands for wastewater treatment: their potential for nutrient removal. No. Transformations of Nutrients in natural and constructed wetlands. Backhuys publishers, Leiden, the Netherlands.

White, M.J., and J.G. Arnold. 2009. Development of a simplistic vegetative filter strip model for sediment and nutrient retention at the field scale. Hydrological Processes 23(11):1602-1616.

Whitehead, P.G., E.J. Wilson, and D. Butterfield. 1998. A semi-distributed Integrated Nitrogen model for multiple source assessment in Catchments (INCA): Part I — model structure and process equations. Science of the Total Environment 210–211(0):547-558.

Williams, J.R., C.A. Jones, and P.T. Dyke. 1984. A modeling approach to determining the relationship between erosion and soil productivity. Transactions of the ASAE 27:129-144.

Williams, J.R., A.D. Nicks, and J.G. Arnold. 1985. Simulator for Water Resources in Rural Basins. Journal of Hydraulic Engineering 111(6):970-986.

Winsemius, H.C., B. Schaefli, A. Montanari, and H.H.G. Savenije. 2009. On the calibration of hydrological models in ungauged basins: A framework for integrating hard and soft hydrological information. Water Resources Research 45(12):W12422.

Yang, Q., F.A. Benoy, T.L. Chow, L.-L. Daigle, C.P.-A. Bourque, and F.-R. Meng. 2011. Using the Soil and Water Assessment Tool to estimate achievable water quality targets through implementation of beneficial management practices in an agricultural watershed. Journal of Environmental Quality 41(1):64-72.

Yang, Q., F.-R. Meng, Z. Zhao, T.L. Chow, G. Benoy, H.W. Rees, and C.P.A. Bourque. 2009. Assessing the impacts of flow diversion terraces on stream water and sediment yields at a watershed level using SWAT model. Agriculture, Ecosystem and Environment 132(1-2):23-31.

Yang, Y., and L. Wang. 2010. A Review of Modelling Tools for Implementation of the EU Water Framework Directive in Handling Diffuse Water Pollution. Water Resource Management 24(9):1819-1843.

Young, R.A., C.A. Onstad, D.D. Bosch, and W.P. Anderson. 1986. AGNPS: A nonpoint-source pollution model for evaluating agricultural watersheds. Journal of Soil and Water Conservation 44(2):168-173.

Youssef, M.A., R.W. Skaggs, G.M. Chescheir, and J.W. Gilliam. 2005. The nitrogen simulation model, DRAINMOD-N II. Transactions of the ASAE 48(2):611-626.

LIST OF TABLES

LIST OF FIGURES

ABBREVIATION

DEM	Digital Elevation Model
GEUS	Geological Survey of Denmark and Greenland
GLUE	Generalized Likelihood Uncertainty Estimation
HRU	Hydrological Response Unit
LU	Landscape Unit
NSE	Nash-Sutcliffe Efficiency
RNM	Riparian Nitrogen Model
SWAT	Soil and Water Assessment Tool
VFS	Vegetative Filter Strips

ABOUT THE AUTHOR

Linh Hoang was born on 31 December, 1983 in Ho Chi Minh city, Vietnam. She did her bachelors degree in Environmental Management in the Faculty of Environment, Ho Chi Minh city University of Technology, Vietnam. She graduated for her Bachelor degree in 2006 with a Gold Medal award for being the first rank among all students in the Faculty of Environment. With that good performance, she was offered to work in her Faculty as a lecturer after her graduation. She started as an assistant lecturer and also joined several projects on water quality monitoring and management and Environmental Impact Assessment. She then won the Huygens scholarship from the Dutch government to pursue her Master studies in Hydroinformatics at UNESCO-IHE Institute for Water Education, Delft, the Netherlands. During her Master thesis she used the Soil and Water Assessment Tool (SWAT) to integrate river basin modelling integrating with water quality modelling. She obtained her MSc degree with Distinction in 2009. With the interest on doing research and the eagerness to explore more about the world, she decided to continue her PhD degree in the same Institute for Water Education, UNESCO-IHE. In her PhD research she continued to use the SWAT model, but focused on improving the model processes by adding the landscape routing process to hydrological modelling and a sub-module to simulate the denitrification in riparian zones in the SWAT model.

PUBLICATIONS

Journals

L. Hoang, A. van Griensven, P. van der Keur, J. C. Refsgaard, L. Troldborg, B. Nilsson and A. Mynett, 2012. Comparison and evaluation of model structures for the simulation of pollution fluxes in a tile-drained river basin. *Journal of Environmental Quality*. doi:10.2134/jeq2011.0398

L. Hoang, A. Mynett, A. van Griensven, Extending the SWAT modelling suite for representing landscape variability and simulating denitrification in riparian zones at the river basin scale (submitted to Hydrological Processes)

L. Hoang, A. Mynett, A. van Griensven. Evaluation of the effect of nitrate removal by denitrification at the river basin scale using the SWAT model. (in preparation)

Conferences

L. Hoang, A. Mynett, A. van Griensven, 2013. Modelling water and nitrogen transport and transformation across the landscape at river basin scale using SWAT. Proc. of 2013 IAHR World Congress, Chengdu, China

L. Hoang, A. van Griensven, A. Mynett, 2013. Modelling the effeciency of nitrate removal by denitrification in the SWAT model, 2013 International SWAT conference. Toulouse, France (oral presentation)

L. Hoang, A. van Griensven, A. Mynett, 2012. A new approach representing landscape variability for the SWAT model. PUB symposium "Completion of the IAHS decade on Prediction in Ungauged Basins and the way ahead", Delft, the Netherlands (poster presentation)

L. Hoang, A. van Griensven, A. Mynett, 2012. Evaluation of different model structures for flow and nitrate in a tile-drained river basin. First European Symposium on Remediation Technologies and their Integration in Water Management, Barcelona, Spain (oral presentation)

L. Hoang, A. van Griensven, A. Mynett, 2011. Simulation of the hydrological and nitrogen balance and cycle within the odense river basin. 3rd International Multidisciplinary conference on Hydrology and Ecology: Ecosystem, Groundwater and Surface Water - Pressures and Options, Vienna, Austria (poster presentation)

L. Hoang, A. van Griensven, A. Mynett, 2011. Simulation of tile drainage in an agricultural catchment in Denmark. Proc. of SWAT-SEA II International conference, Ho Chi Minh city, Vietnam

L. Hoang, A. van Griensven, P. van der Keur, L. Troldborg, B. Nilsson, and A. Mynett. 2010. Comparison of the SWAT model versus the DAISY-MIKE SHE model for simulating the flow and nitrogen processes. Proc. of 2010 International SWAT conference, Seoul, Korea

L. Hoang, A. van Griensven, J. Cools, and A. Mynett. 2009. Instream water quality processes in SWAT with different routing methods and adapted water quality modules for daily or sub-daily time steps. Proc. of 2009 International SWAT conference, Boulder, Colorado, USA

T - #0412 - 101024 - C200 - 240/170/11 - PB - 9781138024052 - Gloss Lamination